RISK MANAGEMENT FOR DESIGN PROFESSIONALS

by William G. Ramroth, Jr., AIA

KAPLAN PUBLISHING
New York

Vice President and Publisher: Maureen McMahon
Editorial Director: Jennifer Farthing
Acquisitions Editor: Michael Sprague
Development Editor: Joshua Martino
Production Editor: Karina Cueto
Cover Designer: Rod Hernandez

1 Liberty Plaza, 24th Floor
New York, NY 10006

Printed in the United States of America

October 2007
07 08 09 10 9 8 7 6 5 4 3 2 1

ISBN 13: 978-1-4277-5476-9

To Dad the engineer, from whom I first learned to respect and love the design professions

The author has attempted to make the information contained in this book as helpful and factual as possible. The reader is advised that nothing in this text is intended to serve as legal or professional advice. Before acting on any of the opinions or recommendations contained herein, the reader should seek legal and professional advice.

Contents

Introduction vii

1. Principles of Risk Management 1

Some Outcomes are More Likely than Others Are 6

Risks and Rewards are not Always Proportional 9

Base Decisions on Logical Conclusions about Potential Rewards 14

The Whole Can Be Known from the Parts 17

Things are Predictable Within a Range 20

Facts and Utility Are Not the Same 24

The Vital Few and the Trivial Many 26

Monte Carlo or "What-If" Modeling 27

Pragmatic Inference and Risk Management 28

2. The Risk Management Process 33

The Steps of Risk Management 35

Determining Overall Risk 46

3. Comprehensive Risk Management 61

Narrow versus Wide View 63

Comprehensive Risk Management 65

The Business Plan 70

Employee Relations and Management Plan 74

Office Reporting and Procedures Plan 77

4. Risk and Business Development 79

Marketing, Business Development, and Sales 81

Approaches that Don't Work in the Long Run 83

Business Development Basics 86

The BD Plan 94

The Long and Short Course of Business Development 100

Go/No Go Form 104

Project Manager's Role in Business Development 110

Risk and Proposal Writing 111

5. Risk and Project Management 117

Project Manager Skills 118

Why Projects Fail 125

Risk Management Techniques 132

Managing Risk While Executing the Project 141

6. Cost, Consultants, and Codes 149
Designing to Construction Budget 150
Controlling Design Budgets 157
Consultants 157
Codes 160

7. Professional Liability and Contractual Risks 173
Product Liability versus Professional Liability 174
Professional Liability Insurance 177
Independent Brokers and Insurance Company Agents 179
Professional Liability Policies 180
Occurrence, Prior Acts, and Claims-Made Insurance Policies 182
The Well-Off and the Not-So-Well-Off 184
Standard of Care 185
Managing Liability Exposure 187
Claims Response Plan 187
Even the Innocent Pay 191
Professional Services Contracts 192
Working for Free 203

8. Risk and Construction Support Services 207
Services During Construction 209
Risk and Design/Build 230

9. Risk Management Rules of Thumb 235
Rule of Thumb No. 1: Know the Territory 236
Rule of Thumb No. 2: The 3.0 Multiplier Rule of Thumb 237
Rule of Thumb No. 3: Manage Scope Creep or it will Break Your Back 239
Rule of Thumb No. 4: Don't Push the Panic Button 241
Rule of Thumb No. 5: The 60 Percent Rule of Thumb 242
Rule of Thumb No. 6: Profitability has Two Main Rudders 244
Rule of Thumb No. 7: Somebody Is Keeping Score 246
Rule of Thumb No. 8: Drawings Are Never 100 Percent Complete 246
Rule of Thumb No. 9: The Hurrieder I Go the Behinder I Get 248
Rule of Thumb No. 10: Document, Document, Document 249

Appendix: Risk Management Checklist 251
Acknowledgments 275
Notes 277
Bibliography 281
Index 285

Introduction

Charles Bulfinch (1763–1844) is considered to be the first professional architect in America.[1] In 1784, he graduated from Harvard University where he studied mathematics and perspective. The following year, he left Boston and sailed to Europe to study architecture. Bulfinch spent time traveling through England, France, and Italy examining, measuring, and drawing the great buildings he saw. He bought books on art and architecture and taught himself how to design. Then, in January 1786, Charles Bulfinch returned to Boston and decided to be an architect.

There was a problem, however. At the time, American architecture was a risky business. Actually, it was no business at all; it was not a recognized profession. Unlike Europe, America was a new world that lacked many of the professions that had existed in Europe for centuries—architecture being one of them. Instead, it was an avocation, a pastime enjoyed by gentleman-amateurs. The architectural historian William H. Pierson, Jr. explained that the late 18th-century American architect was someone:

> ...whose principal vocation might be anything from merchant to college professor and who designed buildings as a condition of his status as a gentleman. He rarely, if ever, played an active role in construction, however, and he was never paid except in the form of some token gift.[2]

In late 18th-century America there were no schools of architecture, no apprenticeship programs, no professional organizations, and no standards regarding the practice and business of design. Most significantly, there was no way to get paid as an architect. Nevertheless, Bulfinch had his mind set on making a living as an architect.

Today, architecture, as well as engineering and planning, are well-established design professions. Their members are highly respected and well paid. Today's design professionals, of which there are hundreds of thousands, have generations of professional precedence upon which to rely and build. Universities and colleges now have nationally accredited programs with standardized curricula for teaching the design professions, and there are hundreds of design firms of all types and sizes. Professional organizations, apprenticeship and training programs, and state licensing requirements now exist. There are even insurance underwriters and attorneys who specialize in servicing and advising the design professions. Established business and marketing practices have evolved, and there are accepted professional practices, standards of ethics, and various contracts and methods of payment, all with proven track records.

Despite all its advances, design is still a risky business. There are a myriad of things that can go wrong if they're not anticipated and properly managed. To assist in managing these risks, the modern design professional has a variety of tools and resources available, not the least of which are the lessons learned from the successes and failures of others.

Bulfinch had no such luxury. He charted the risky seas of a fledgling profession all by himself and learned the hard way. He began his career with a flurry of work, yet within a few years, he was bankrupt.

Charles Bulfinch was ambitious and his family was affluent, influential, and highly respected throughout Boston. He had little trouble landing his first nonpaying jobs—the designs of the Hollis Street Church, a theater on Federal Street, and a state house for Massachusetts. By November 1787, less than a year after his return from Europe, he submitted his first design drawings for the Massachusetts State House.[3]

For the design of the Federal Street Theater, Bulfinch received no commission. His design work was considered a gift. In exchange, Bulfinch received a gift, too. His granddaughter and first biographer, Ellen Susan Bulfinch, wrote:

> In recognition of the gift from the architect of the plans for the building, a gold medal, of much beauty of workmanship, was presented to Mr. Bulfinch by the owners...[4]

While working on the design of the Federal Street Theater and the Massachusetts State House, Bulfinch undertook another project. This one would be a paying job, or at least that's what Bulfinch thought. He entered into a tontine partnership with two others—a contractual arrangement among partners that specifies that if one partner dies, the others agree to pay off the deceased partners' survivors, dividing the shares equally, and so on until there is only one partner left.

Bulfinch and his tontine partners' scheme was simple and has been repeated by experienced and inexperienced developers thousands of times since with mixed results. The tontine would buy a vacant lot, build houses, sell the houses, and split the profit. Bulfinch would be the designer and he would get paid for his design work, albeit not strictly as an architect.

The partners bought land on Franklin Street near the Franklin Street Theater that Bulfinch had designed. Bulfinch designed a gracefully curved, crescent-shaped 400-foot-long grouping of 16 townhouses. The original plan included two such crescent-shaped structures, totaling 32 townhouses, with an elongated oval park in the center. Only half of the project was ever built. Originally called Franklin Place, today architectural historians refer to it as the Tontine Crescent.

Bulfinch based its neoclassical design on crescent-shaped buildings he had seen in Paris and the Royal Crescent in Bath, England, designed by John Wood. In addition, while in London,

England, Bulfinch had seen drawings by the Scottish-born architect Robert Adam for crescent-shaped Portland Place.

When the first tontine crescent was complete, many of the units were still not sold. One of Bulfinch's partners had second thoughts about his investment and wanted out. Because the other partner was financially strapped, Bulfinch agreed to buy out the skittish partner, borrowing heavily at great risk, and compounding it by getting his father and brother to cosign the loans.

Unfortunately, Boston entered into a severe real estate recession. The remaining townhouses did not sell and Bulfinch lost everything. Later in his life, Bulfinch began to write an autobiography, which he never finished. In it he expressed his regret over the enormous risk he had taken:

> [No further] sales could be made and no further loans obtained and I was obliged to become bankrupt and assign the property for creditors; in the consequences of this project, my father and brother…were involved as my endorsers. With what remorse have I looked back on these events when blindly gratifying a taste for a favorite pursuit, I involved for life myself and wife with our children—my Father and Mother and sisters, who all held the utmost confidence in my measures and pride in my expected success.(5)

The "favorite pursuit" that Bulfinch mentioned was architecture. It was his love of architecture and desire to be an architect that affected his judgment and led him to take the uncalculated risk that he did.

For design professionals, there is a risk management lesson in Bulfinch's story. Elsewhere in his autobiography Bulfinch touched on it. Writing about his grand tour of Europe and self-education as an architect, Bulfinch opined:

> I was delighted in observing the numerous objects & beauties of nature & art that I met with on all sides, particularly the wonders of Architecture, & the kindred arts of painting & sculpture...but these pursuits did not confirm me in any business habits...on the contrary they had a powerful adverse influence on my whole after life.[(6)]

Today design is a business as well as an art, science, and profession. It is important for design professionals to recognize this basic truth and understand the many risks that go along with practicing a profession that is simultaneously a business.

Few students choose design because it is a business. Most choose one of the design professions because they have an interest in the art and science of architecture, engineering, or planning, not because they like or understand the practice and business of design. In general, design professionals do not come from business schools, although some have studied business. They come from schools of architecture, engineering, and planning. For most design professionals, what they know about the practice and business of design has come from on-the-job learning. Sometimes these lessons were learned the hard way.

The purpose of this book is to help the design professional recognize and manage the many threats, dangers, and undesirable events that adversely affect the practice and business of design. The word "risk" is commonly used to stand for the possibility of a threat, danger, or undesirable event occurring.

Risk is a fact of life. There is always the possibility of an undesirable event occurring. Just getting out of bed in the morning has risks as does stepping into the shower, or walking to the bus stop on the way to work. Something dreadful can happen at any time. For most daily activities, however, the possibility of something undesirable occurring is minor. It can be ignored or easily controlled by doing what comes naturally.

The design business, however, is not one of these activities. It is not simple. Much of the process and business of designing the built environment is complicated. Some risks have high probabilities of occurring if not properly managed. The practice of design is not exactly like playing roulette or throwing dice, but the design firm that does not manage risk is definitely playing a game of chance. Sooner or later the odds are good that some untamed risk will rear its ugly head and cause problems, perhaps even jeopardizing the firm's existence.

"Risk Management" is the art and science of recognizing, preventing, and mitigating threats and uncertainties. Risk management assumes that possible threats or dangers are identifiable in advance, measurable—at least qualitatively—and preventable or controllable.

Risk Management for Design Professionals examines the many risks that affect the practices of architecture, engineering, and planning, and suggests ways of managing them. This book discusses the risks associated with business development, design, project management, production of construction documents, and in providing construction support services. It examines the various risks associated with typical owner/design professional agreements, identifying clauses, phrases, and words that increase and decrease risk. It discusses how the various types of contracts between owners and constructors impart different levels of risk to design professionals. *Risk Management for Design Professionals* discusses how to select and work with consultants to minimize risk. The text takes a look at code and zoning requirements and suggests ways of managing the risks associated with planning and building department review processes.

This book's first two chapters give an overview of the basic principles and process of risk management. 3 explains the importance of developing and implementing a comprehensive risk management plan. Chapters 4 through 8 look at the risks associated with business development, project management, design, preparation of construction documents, building and zoning code reviews, working with consultants, agreements, professional liability insurance, and providing construction support services.

FIGURE A *East exterior elevation of the U.S. Capitol Building, Washington, D.C., showing Bulfinch's great wooden dome completed in 1829. This daguerreotype (an early form of photography) is believed to have been taken by John Plumbe, Jr. around 1846, and is the earliest known photographic image of the Capitol. © Bettmann/CORBIS*

9 includes risk management rules of thumb. A risk management checklist is included in the Appendix. It serves as a summary and quick reference for the risk management topics discussed in the book. Throughout the book, historical, contemporary, and hypothetical examples are used to explain and illustrate risk and risk management.

It would be disappointing to leave the introduction with Bulfinch in a state of bankruptcy. Indeed, the story of the first professional architect in America has a happy ending. Soon after declaring bankruptcy, Bulfinch, in desperate need of income, took the job of Boston's Superintendent of Police, which paid $600 per year. Bulfinch wrote about his job:

> Although this employment was irksome and little suited to my taste or character, I undertook to discharge its duties, and it enabled me, with my personal attention

> as Architect, to support my family in respectable simplicity, which my dear wife submitted to with cheerfulness.[7]

Bulfinch continued to design buildings and oversee their construction. In time, he became the most active architect in Boston. In addition, he was instrumental in improving Boston's streets, street lighting, and storm water and sewage drainage systems. His reputation as an architect and city planner grew and in 1818, he moved to Washington, D.C., where he took the position of Commissioner of Public Buildings, with a handsome salary of $2,500 per year. While holding the post of commissioner, Bulfinch succeeded Benjamin Latrobe as architect for the United States Capitol, which the British had burned in 1814. Bulfinch designed and oversaw the construction of the Capitol's portico, wings, and central portion, which he enclosed under a great wooden dome. (See Figure A.) The Capitol was completed in 1829. When Bulfinch died in 1844, his reputation as an important and influential American architect was well secured.

Ultimately risk management decisions are judgments regarding two or more possible alternatives. The judgments can be made instantly, without regard to thought or facts, or they can be made methodically, based on reasoned processes, data, probabilities, experience, and sound decision-making principles. The latter method is much more likely to yield consistently desirable outcomes. Therefore, it is to the decision-making principles that form the basis of sound risk management decisions to which we first turn.

Chapter

1

PRINCIPLES OF RISK MANAGEMENT

Engineer Dankmar Adler and architect Louis Sullivan stood in the doorway that separated two ages of architecture. Behind them were four centuries of classical architecture, with a style so well entrenched that it is also known as traditional architecture. Ahead of them lay the uncharted age of modern architecture and the skyscraper. Together, they helped push architecture over the threshold and into the Modern Age.

Adler and Sullivan were not opposed to experimentation. Nevertheless, they did not take unnecessary risks with the design and construction of their buildings. Nor did they rely on luck to make their projects successful. Through deliberate actions, they managed the risks associated with their pioneering architectural and engineering work.

Adler and Sullivan's Auditorium Building in Chicago, Illinois, was designed and built between 1886 and 1890. Part theater, part office building, and part hotel, the Auditorium Building was one of the largest and most complex multi-use buildings of its time. When completed, it catapulted Adler and Sullivan, as well as the city of Chicago, onto the architectural world stage.

The design and construction of all buildings involves some risk, and revolutionary works, such as the Auditorium Building, involve even more. A key to success in the design professions is to manage the risks that can potentially threaten a project. This requires identifying risks, analyzing them, prioritizing their severity, and developing a plan to control or mitigate the risks that pose the greatest threat. Today, we call this process "risk management." Although Adler and Sullivan probably never heard the term, they, nevertheless, employed risk management during the design and construction of the Auditorium Building.

During the 1880s, hundreds of buildings were built in Chicago. The heavier ones had a tendency to sink into the Lake Michigan shoreline muck that lay beneath the city. The problem was not new to Chicago—the risk of building in poor soil has been around for millennia. In 30 B.C., the ancient Roman architect Vitruvius recommended a method for reducing the risk:

> [F]oundations...should be dug out of the solid ground...and carried down...as far as the magnitude of the work shall seem to require, and the whole substructure should be as solid as it can possibly be laid...If, however, solid ground cannot be found, but the place proves to be nothing but a heap of loose earth to the very bottom, or a marsh, then it must be dug up and cleared out and set with piles made of charred alder or olive wood or oak, and these must be driven down by machinery, very closely together...and the intervals between them filled in with charcoal.[8]

A particularly serious problem is differential settlement. If different sections of a building sink faster or deeper than others do, foundations, slabs, and building materials can crack. Extreme differential settlement can jeopardize the very stability of a structure. The Leaning Tower of Pisa is perhaps the world's best-

FIGURE 1.1 *Southwest view showing the 17-story tower of the Auditorium Building extending seven stories above the main portion of the building. Auditorium Building, 430 S. Michigan Avenue, Chicago. Adler and Sullivan, architects. Built 1887–1899. View c. 1890. Barnum and Barnum, photographers. Historic Architecture and Landscape Image Collection, Ryerson and Burnham Archives, The Art Institute of Chicago, Digital File © The Art Institute of Chicago.*

known example of what can happen when a building is subjected to severe differential settlement.

The Auditorium Building was an extremely heavy masonry bearing wall building, weighing over 100,000 tons, setting on mud. Its footprint was 63,350 square feet (about an acre and a half). It was 10 stories tall, except for its tower, which continued up an additional seven stories to a height of 275 feet (see Figure 1.1).

Adler was particularly worried about differential settlement. To reduce the risk, he designed the foundation like a giant raft. This helped to distribute the enormous weight of the building evenly over the soil beneath it. Approximately 28,000 cubic yards of shoreline mud was removed to construct the foundation and basement for the Auditorium Building. Alternating courses of pine timbers, concrete, and steel rails were installed to form the building's raft-like foundation.

Generally speaking, as a building is constructed, its weight gradually and evenly increases. However, the tower portion of the Auditorium Building created a special problem. Because it was seven stories taller than the rest of the building, the foundation beneath the tower would not be loaded at the same rate as the rest of the building's foundation, thus increasing the risk of differential settlement during construction. To mitigate the risk, Adler came up with an ingenious idea. It so elated Sullivan that he called Adler's idea a "regular Columbus egg stunt."[(9)]

Adler's solution was to preload the tower's foundations in proportion to the weight of the rest of the building. As the building grew taller, Adler calculated the additional weight required to load the foundation beneath the tower uniformly and then stacked an equal weight of bricks and pig iron in the tower's basement and lower floors. Periodically he checked the building for cracks. Once the tower portion reached 10 stories—the height of the rest of the building—Adler began removing the bricks and pig iron in proportion to the weight of the growing tower. By the time the tower topped out at 17 stories, all the bricks and pig iron had been removed. The result was that the building was constructed without cracks; the risk of differential settlement was successfully avoided.

The Auditorium Building still stands at 430 S. Michigan Avenue. While the building has settled approximately 18 inches over the course of 125 years, it is nearly free of cracks due to differential settlement, a testimony to Adler's skill and risk management planning.

Adler's solution for controlling differential settlement illustrates the six basic steps of the risk management process:

1. Create a risk management plan
2. Determine the risks

3. Prioritize the risks
4. Determine which have the greatest impact
5. Develop solutions
6. Implement the solutions

The first step is to understand that every project has risks and that every project needs a plan to manage the risks. Second, determine the risks that can adversely affect the project's success. Next, prioritize the risks based on their likelihood (the probability that they will occur), and then determine which risks have the greatest severity of impact on the project. Then, develop solutions that reduce or eliminate the occurrence of the risks that can most jeopardize the project's success, and finally, implement those solutions or procedures that reduce or eliminate the risks that pose the greatest threat and monitor their effectiveness. Chapter 2 discusses these six steps in greater detail.

As the story about the Auditorium Building shows, there is nothing new or mystical about risk management. Without using the term or being aware of the six steps, Adler was actively engaged in risk management. The difference between Adler and Sullivan's time and ours is our recognition of the process and use of the term. Since Adler and Sullivan's era, much study has gone into defining and refining the principles of risk management.

The discussion that follows is an overview of the basic tools and principles required to manage risk. The tools are not tools in the literal sense, such as hammers or screwdrivers. They are intellectual tools, tools of logic and reason. In addition, the text provides a cursory look at their origins.

We begin with an essential tool we take for granted: fractions, or as we often express them, percentages. The use of fractions to express probability is perhaps the most fundamental tool of risk management. In addition, the use of fractions quickly leads to another fundamental principle of risk management: Some outcomes are more likely than others are.

SOME OUTCOMES ARE MORE LIKELY THAN OTHERS ARE

Many years ago I had a conversation with the principal of a moderately sized engineering firm. His firm responded to a request for qualifications and had been short-listed for a project, along with three other firms. Since his firm was one of four firms being considered for the job, he quickly did a little arithmetic and told me that, "We have a one-in-four [1/4] chance of winning the job."

Today, fractions are used all of the time to express probability. They are used to express the ratio of two whole numbers—the outcome desired or in question divided by the total number of possible outcomes. Sometimes the fraction is expressed as a percentage—so many outcomes or chances out of 100.

The use of fractions to express probability is not as old as you might think. Civilization got along for millennia without using fractions to quantify the likelihood of something happening.

The first known recorded use of fractions to express possible outcomes dates from the 16th century and the Italian Renaissance. They were used by an incorrigible gambler who was, no doubt, interested in improving his chances of winning, while at the same time reducing his chances of losing.

Girolamo Cardano (1501–1576) was an Italian physician, mathematician, and gambler. He gambled nearly every day, betting on dice, cards, and even chess. In his autobiography *De vita Popria Liber (The Book of My Life),* Cardano confessed to applying himself assiduously to gambling for almost 40 years.[(10)]

Cardano was also a prolific writer and published 131 works in his lifetime. He had another 111 in manuscript form at the time of his death. Among the unpublished manuscripts was a work entitled *Liber de Ludo Aleae* or *The Book on Games of Chance.* The book was not published until 1663, well after his death. In his autobiography, Cardano said he first wrote the manuscript in 1525, and then rewrote it in 1565. If the former date is true, then

The Book on Games of Chance is the earliest known work to delve into the principles of probability and use fractions to express possible outcomes.

Like Cardano, the ancient Greeks and Romans loved to gamble. Some gambled away their entire possessions and a few even gambled themselves into slavery. They had no quantitative way to assess their chances. All they had was a rather useless qualitative way. Aristotle summed it up:

> To succeed in many things, or many times, is difficult; for instance, to repeat the same throw ten thousand times with the dice would be impossible, whereas to make it once or twice is comparatively easy.[11]

The ancient Greeks and Romans played a dice game similar to modern-day craps. (Craps, incidentally, is an American game, invented in the early 19th century.) Instead of rolling dice, ancient Greeks and Romans rolled bones called *astralgai* and bet on the outcomes. Astralgai were made from the foot bones of hoofed animals, such as goats or sheep, and they were oblong, not cubical, in shape. Because of their geometry, astralgai could not land on their ends. Consequently there were only four possible ways for astralgai to land. Numbers or characters were inscribed on the four sides. Astralgai were wider on two sides and so the chance they would come to rest on the wider sides was greater than the narrower sides. The ancients seemed not to understand this, for payouts for different combinations were not predicated on the likelihood of the different outcomes.

By the 16th century, cube-shaped dice with equal sides replaced astralgai in many games of chance. Cardano devised a way to determine mathematically the likelihood of various rolls of dice. He did so by using fractions to express the number of favorable outcomes compared to the total number of possible outcomes. For example, the chance of a tossed coin landing heads-up is 1/2 (one chance in two possible outcome), and a

Sum of Two Dice	Probability (expressed as a fraction)
2	1/36
3	2/36 or 1/18
4	3/36 or 1/12
5	4/36 or 1/9
6	5/36
7	6/36 or 1/6
8	5/36
9	4/36 or 1/9
10	3/36 or 1/12
11	2/36 or 1/18
12	1/36

FIGURE 1.2 *Probabilities of sums when rolling two dice.*

rolled die coming up 4 has a chance of 1/6 (one chance out of six possible outcomes).

Using fractions, Cardano was also able to calculate combinations of probabilities. For example, because the probability of throwing a 4 with one die is 1/6 and the probability of throwing a 5 is also 1/6, then the probability of throwing either a 4 or 5 is: $1/6 + 1/6 = 2/6$ or 1/3. The probability of throwing a 4 two times in a row with one die is: $1/6 \times 1/6$ or 1/36.

Today, Cardano's method of dividing the desired outcome by the total number of all possible outcomes is used every day—and not only by gamblers. Businesses use fractions to determine the likelihood of various outcomes. Design professionals also use fractions, for example, to express the chances of winning a job.

Cardano's method of calculating probabilities illustrates a fundamental principle of risk management that can be explained by looking at Figure 1.2, a table that lists the probabilities of the possible sums when rolling two dice.

You will notice that not all possible outcomes are equally probable. For example, the probability of rolling a 2 (snake-eyes) or a 12 (boxcars) with two dice is 1/36. Both are the least likely outcomes compared to all others. Note that the probability of throwing a 7 is 1/6 or six times more likely than throwing boxcars, which explains why the number 7 is so important in the game of craps. Craps players live and die by these odds; consequently, hardcore dice players memorize them.

Risk managers are not gamblers. Unlike gamblers who must live by the odds, risk managers manipulate them. Through deliberate action, they work to make desired outcomes the more likely outcomes. As the chapters that follow show, there are many actions that design professionals can take to skew the odds in their favor and make the outcomes that they want more likely.

RISKS AND REWARDS ARE NOT ALWAYS PROPORTIONAL

Tiger Woods is a terrific golfer and frequently favored to win. During the year 2000, Woods was playing so well that one bookmaker offered odds of only 1-to-2 for Woods to win one of the four major golf tournaments that year. In betting parlance, odds of 1-to-2 means that a winning $2 bet pays back the original $2 plus just $1 in profit—little reward for little risk.

In contrast, the golfer Paul Lawrie was a 200-to-1 favorite to win the 2000 British Open, meaning that if Lawrie won, a $1 bet would yield a $200 return—a big reward for a big risk or, said another way, a big payoff for an unlikely outcome. Ironically, Lawrie had won the British Open the year before. In 1999, golf enthusiasts all over the world watched in disbelief as leader Jean Van de Velde blew up on the final hole of the British Open, handing Lawrie the victory and sealing Van de Velde's final-hole collapse in the annals of golf as one of the biggest chokes in golfing history. Lawrie won because of a fluke, and odds-makers don't base odds on flukes. As it turned out, Woods won the British Open in 2000, and his win paid $6 on a $1 bet. In general, "risk to reward" works as follows: Little risk yields little reward and bigger risks yield bigger rewards.

However, not all risks to rewards are proportional. Sometimes, big risks offer little reward and little risks offer big rewards. Two examples from the 17th century illustrate these extremes. The first involves a flamboyant, rich, hard-core gambler and the second a fiercely introspective man who explained a very unusual and famous wager.

The French nobleman, the Chevalier de Méré, was well known in gambling houses for his unorthodox habit of making the same bet over and over and over. He bet he could roll a 6 with one die in four tries. Because the bet seemed foolhardy to other gamblers, he had a lot of takers. The Chevalier lost a lot of money, but because he was rich he could afford to keep at it. Over the course of very many bets and his willingness to risk large sums, Méré did come out ahead, as he knew he would. Unknown to his gambling opponents, Méré had a trick up his sleeve: fractions and a bit of algebra.

Recall from the discussion of Girolamo Cardano that his manuscript *The Book on Games of Chance* was not published until 1663. Between the time that Cardano wrote his manuscript and its publication, a few others had also figured out the benefits of using fractions and algebra to express and determine probabilities. The best known among them were the Italian scientist Galileo, the Dutch mathematician and astronomer Christiaan Huygens, and the French mathematician Blaise Pascal. Although he wasn't an intellectual counterpart of Galileo or Huygens or Pascal, Méré was also a mathematician, and he applied math to his gambling passion. Méré figured out that the chance of throwing a 6 in four tries is slightly better than 1/2. To be precise, it is 51.77/100 or 51.77 percent. The odds of throwing a 6 in four tries can be calculated by assuming that the chances of not throwing a 6 in four tries is 5/6 × 5/6 × 5/6 × 5/6 or 0.4823 or 48.23 percent. This makes the chances of throwing a 6 equal to 100 percent minus 48.23 percent, or 51.77 percent.

Méré was financially able to lose almost as often as he won. Because he was willing to slog it out over the long haul, he eventually came out marginally ahead, thus taking a big risk for little reward.

His approach poses an interesting question for design professionals. How many design firms could afford to use Méré's strategy? How many are willing to go over budget on nearly half their projects in hopes that profit from the successful half will more than offset the losses from nearly an equal number of clinkers? Using Méré's approach, how long could most design firms afford to keep their doors open?

The answer is that the vast majority of design firms would not last long. The lesson from Méré for design professionals is that taking large risks for little reward is no way to run a design business. It behooves design professionals to recognize actions with enormous risks and steer clear of them.

There is an interesting postscript to the Méré story. Eventually he grew bored of making the same bet over and over, so he decided to try a new one. He bet he could roll double sixes on 24 rolls of two dice. Méré's mathematical skills were limited and on this bet he miscalculated. The odds of rolling boxcars in 24 tries are 49.14 percent—less than 1/2. Eventually Méré figured this out the hard way.

History's source for Méré's mathematical prowess (or lack of it) is his friend Blaise Pascal. Pascal (1623–1662) was a French mathematician and philosopher. As a child, he excelled in mathematics. At age 16, he wrote an essay on geometry and conic sections. At 18, he invented the first calculating machine, and by his mid-20s he was engaged in the study of probability and games of chance through the exchange of letters with Méré and another French mathematician, Pierre de Fermat, 20 years his elder. In one letter to Fermat, Pascal commented on the math skills of their mutual friend Méré, "He's a fun guy but, alas, no mathematician."[(12)]

Pascal, on the other hand, was a brilliant mathematician. He wrote that "mathematical laws must dominate the wishes of the players [of games of chance] themselves," and that players "are only abstractions of a general principle."[(13)] Pascal's method of solving mathematical problems was logical, rational, calculating, and detached from the desires of the players. As a philosopher and fiercely religious man, however, Pascal's approach to problem solving was a bit different. Here, subjectivity played an important role in his reasoning.

In his book *Pensées,* Pascal combined his interests in probability, games of chance, and religion into a usual wager. He posed the question, is it better to believe in God or not to believe in God? His thought-experiment has come to be called Pascal's Wager and it has been heralded as an excellent example of risk management when the future is unknown. Pascal argued that it is not possible to know whether God exists or not

> but to which view shall we be inclined? Reason cannot decide this question. Infinite chaos separates us. At the far end of this infinite distance a coin is being spun which will come down heads or tails. How will you wager? Reason cannot make you choose either, reason cannot prove either wrong.[(14)]

Nevertheless, according to Pascal, we must choose because we are already committed to the game of life and at stake are our very souls. Note that in the previous quotation that Pascal set the odds at 50 percent that God exists—"heads or tails," he called it. Pascal concluded that a wager that God exists is favorable to a wager that He does not. Pascal set up his argument as a series of premises followed by their logically derived conclusions. His reasoning went something like this: The atheist believes that God does not exist and that all that exists is this world. For the atheist, there is nothing after death, no God, no judgment, no heaven, and no hell. If the atheist is right, there is nothing at risk, nothing to gain or lose. Life ends with death. If, however, the atheist is wrong there is an infinite risk. If God exists then there is His judgment and, for the atheist, there will be eternal hell to pay. For the believer, God exists and there is life after death. If the believer is correct, living a good and moral life on earth yields an infinite reward, salvation, and eternal bliss in heaven. If the believer is wrong, the believer's worse case is no worse than the atheist's best case, which is that there is nothing after death. In Pascal's view, a wager that God exists is, therefore, logically superior and "carries infinite weight when… there are even chances of winning and losing and an infinite prize to be won."[(15)]

For design professionals, the point of Pascal's Wager is not the validity of Pascal's premises or his specific conclusion, but rather the logic of the decision-making process. Pascal's Wager illustrates two important risk management principles: 1) very minor risks can yield great rewards, and 2) base decisions on logical conclusions about potential rewards. From Pascal's point of view, there is virtually no risk in believing in God, but the reward in doing so is potentially infinite.

For design professionals, there are often projects and business opportunities that carry with them very little risk yet have significant potential rewards. There are also those that are the opposite: great risks with little reward. The trick is to recognize the difference, pursue the opportunities with greater rewards, and avoid those that have greater risks, unless the design professional is willing to live with the possible dire consequences.

A good example of this is marketing additional design services to an existing client. It generally carries with it few additional business development costs, yet has the potential of significantly increasing the amount of design fees from that particular client. Another example is the implementation of a quality assurance program that requires all construction documents to be checked before they leave the office. Once a quality assurance program is established, all construction documents can be consistently and thoroughly checked following set procedures with only minor additional production costs. Checking and correcting work before it leaves the office has enormous benefits because it increases the accuracy of the documents and thereby significantly reduces the risks associated with documents of poor quality. Minor risk in the form of a small increase in production costs yields great rewards.

However, the other principle derived from Pascal's Wager is still more significant.

BASE DECISIONS ON LOGICAL CONCLUSIONS ABOUT POTENTIAL REWARDS

Imagine how much easier it would be to make decisions if we knew the future as well as we know the past. Unfortunately, there are always uncertainties about the future. For some design professionals, decision making in the face of uncertainty is paralyzing. Nevertheless, many decisions must be made without knowing for certain which decisions are absolutely the best.

Pascal's Wager presents a logical method for analyzing competing alternative decisions when the future is uncertain. We must draw logical conclusions about probable outcomes from accurate and reasoned premises. If confronted with probable outcomes of equal uncertainty, direct your actions toward the outcome with the greater potential reward or lesser penalty should it eventually prove to be the wrong choice.

Ian Hacking, philosophy and history professor at the Collège de France, claimed in his 1975 book *The Emergence of Probability: A Philosophical Study of Early Ideas about Probability Induction and Statistical Inference* that Pascal's Wager offers a thoroughly modern method for decision making. Pascal's Wager marked the beginning of decision-theory, which Hacking defined as "the theory of deciding what to do when it is uncertain what will happen."[16]

In his wager, Pascal presented the wager in the form of premises. Based on the premises, Pascal drew relevant and persuasive conclusions that logically followed from his premises. According to Peter L. Bernstein, economic and risk management consultant and author of *Against the Gods: The Remarkable Story of Risk,* logically deciding what to do when the future is uncertain "is the essential first step in any effort to manage risk."[17]

Of course, design professionals are not risking their souls with the decisions they make, but they are frequently risking their time, money, and reputation (and that of their clients). Also at stake are the health and safety of the general public that uses the facilities created by design professionals. Nevertheless, for design professionals, the decision-making process is fundamentally the same as Pascal's:

1. Establish the premises or factors that bear on the decision.
2. Identify possible alternative decisions that satisfy the premises or factors.
3. Analyze the merits of the alternative decisions, considering all relevant data making sure to consider the risks involved among competing alternatives.
4. Determine the likelihood of the risks occurring.
5. Choose the decision that logically offers the best solution considering the relevant data and risks involved.
6. Make the decision and know that the decision contains uncertainty because decisions about the future always do.

These are the basic steps in sound decision making. Sound decisions reduce risks because they are logical results of the premises, are based on relevant data, and are an honest assessment and appreciation of the risks involved.

Suppose the following confront a design firm: There are two equally interesting projects worth pursuing with equal probabilities of winning. One has a potential $100,000 fee and the other has a $1 million dollar fee. The design firm must choose which one to pursue. Given that all design professionals have a finite amount of time, capital, and resources (available staff) this is not an unusual choice. Often design firms must choose which projects to chase and which to let go.

Based on a cursory reading of Pascal's Wager, the answer seems obvious: Choose the project with the greater prize to be won—pursue the $1 million dollar project. For design firms with past experience of successfully designing and managing large projects, the decision may be just this obvious. It often doesn't cost much more to pursue a project with a $1 million fee compared to a $100,000 fee. Chasing the $1 million project is a logical choice because it carries with it no more risk than pursing the smaller project.

However, this cursory reading can be misleading. Remember, it is the process in Pascal's Wager that is important, not necessarily his conclusion. The process requires considering all relevant factors that bear on the decision. For design firms with limited or no experience successfully completing large projects, choosing to pursue the larger project might be the wrong decision. For a firm with no experience completing large projects, a tenfold leap in project size might create project management, staffing, production, budget, and quality control problems that outweigh any benefit.

For many questions facing design professionals, there is no certainty that the decision will be the best decision. While experience is helpful in decision making, design professionals cannot rely solely on the past for decisions about the future. They must also manage the odds, make judgments, and set courses that have greater rewards and fewer downsides, should the future unfold in an undesirable way. Pascal's Wager presents a logical way to make these risk management decisions.

THE WHOLE CAN BE KNOWN FROM THE PARTS

Suppose we have a jar filled with black and red jelly beans. A thick paper is wrapped around the glass, preventing us from seeing inside. Without dumping all the jelly beans out and counting them, is it possible to know the ratio of red to black jelly beans? The 17th-century mathematician Jacob Bernoulli (1654–1705) said yes.

When Jacob Bernoulli died, he left a nearly completed manuscript that his nephew Nicolaus Bernoulli finished and published in 1713. Called *The Art of Conjecturing,* it explained a procedure that can answer the question about the ratio of red to black jelly beans. Bernoulli suggested that we sample the jar. We are asked to take out one jelly bean at a time, record its color, replace it, and draw another. If we do this enough times a pattern will form and it will be the ratio of red to black jelly beans.

To illustrate, let's consider what the chance is that the first jelly bean we pull out will be a red one. The common first response is that there is a 50/50 or 1/2 chance that it will be red. Most likely, however, this would be wrong because there is only a 1/2 chance if there is an equal number of red and black jelly beans. Jars of jelly beans—and the real world for that matter—are random and complicated. However, following Bernoulli's sampling technique we can eventually "almost" know the ratio from the pattern that emerges. We can't be absolutely certain, but we can get very close to certainty, and based on this closeness to certainty we can draw a reasonable conclusion. Bernoulli claimed that this is the basis of rational conjecture:

> We define the art of conjecture... as the art of measuring the probabilities of things as exactly as possible, to the end that, in our judgments and actions, we may always choose or follow that which has been found to be better, more satisfactory, safer, or more carefully considered. On

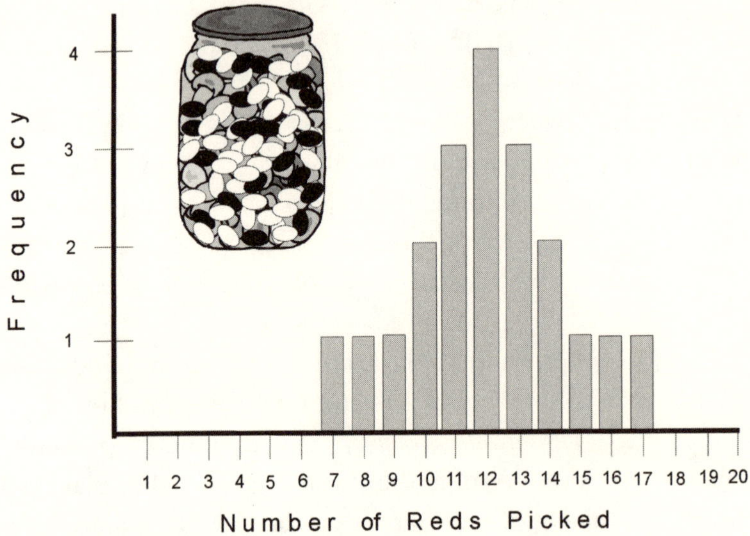

FIGURE 1.3 *Histogram showing the distribution of 20 samplings of red jelly beans.*

> this alone turns all the wisdom of the philosopher and all the practical judgment of the statesman.[18]

We might add to Bernoulli's list, "and the wise decision making of a risk manager."

Note Bernoulli's phrase "as exactly as possible the probabilities of things." He does not say we can know for certain. The only way to be certain is to dump the jelly beans onto the table and count them. However, not everything in the real world is as easy to count as jelly beans. The significance of Bernoulli's approach is that it allows us to know "as exactly as possible" the whole of anything from its parts and, from knowledge about the parts, make decisions.

Returning to our example, how many jelly beans would we have to take from the jar in order to know the red to black ratio as exactly as possible? Moreover, what does "exactly as possible" mean anyway? Bernoulli defined "exactly as possible certainty" as practical certainty, meaning it is as close to certain as can be

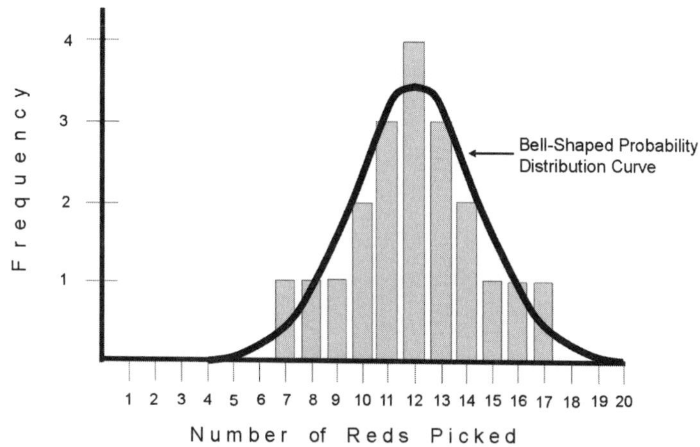

FIGURE 1.4 *A curve superimposed on a histogram showing the distribution of 20 samplings of red jelly beans. Note the bell-shaped form of the curve.*

reasonably attained. He suggested that a certainty of 99.9 percent that an outcome is 98 percent certain is certain enough. Bernoulli called this degree of practical certainty "moral certainty."

His answer to the question regarding how many jelly beans we must take from the jar, however, is not so practical. Bernoulli's calculations showed that if our jar has 5,000 jelly beans in it, we would have to take 25,550 samples in order to reach moral certainty.[19] We would be better off to dump all the jelly beans out and count them one by one! Yet it is generally impossible to count all the parts that bear on a decision. Consequently, Bernoulli's answer to the first question is not very helpful. Nobody is going to take the time to do all the sampling Bernoulli suggested to make a morally certain decision.

A more practical sampling approach is needed, and that approach came along approximately 20 years after the publication of *The Art of Conjecturing*. The next section discusses this approach, but first it is worth summing up Bernoulli's contribution to risk management. He established the principle of practical certainty. To achieve practical certainty, Bernoulli's approach called for the gathering of relevant information (data) and drawing conclusions based on it. He wrote that in order to make a decision based on the likelihood of something occurring, "it is necessary only to calculate exactly the number of possible cases, and then to determine how much more likely it is that one case will occur than another."[(20)] Finally, Bernoulli showed us that the whole can be known by sampling its parts.

There is an interesting story about Bernoulli that illustrates this last principle. The story goes that one day Bernoulli read an anonymous solution to a mathematical problem. By the solution and the way it was presented, Bernoulli knew it had to be the work of the greatest scientist of his time, Sir Isaac Newton. It is said that Bernoulli commented, "I recognize the lion by his paw."[(21)]

THINGS ARE PREDICTABLE WITHIN A RANGE

The French mathematician Abraham De Moivre (1667–1754) read Bernoulli's *The Art of Conjecturing* and wondered if there might be a better way to determine the whole without having to sample so many parts. In 1734, De Moivre published an article that he republished in 1738 in a book on probability called *The Doctrine of Chances.* He presented a method that greatly reduced the number of samples needed to determine practical certainty.

Using De Moivre's sampling approach, we remove 20 jelly beans. Let's say that 12 out of 20 (12/20), or 60 percent, are red. We then return them to the jar, shake it, and remove 20 more. This time 13 out of 20, or 65 percent, are red. We do this a total of 20 times and arrange the results as shown in Figure 1.3 on page 18.

The graphical arrangement of data shown in Figure 1.3 is called a *histogram,* a term first coined by the early 20th-century British mathematician and statistician Karl Pearson. A histogram is a visual aid to a statistical approximation, characterized by columns showing areas of frequency corresponding to units or ranges defined along the base of the histogram. Note in Figure 1.3 that the samples are distributed so that the percentage outcome with the greatest frequency is centered and those of lesser frequency are arranged around it in order of descending frequency. Figure 1.4 on page 19 shows the same histogram with a superimposed curve.

De Moivre showed that the outcomes (in our case the number of red jelly beans) naturally distribute themselves along a curve with the largest number of like observations clustered in the middle, called the *mean,* with steep sides that turn into gradually sloping lines. The plotted data forms a symmetrically shaped curve, like the one shown in Figure 1.4. In 1872, the French mathematician Esprit Pascal Jouffret likened the curve to the shape of a bell. Today, we call the graphical representation of a distribution of samples or outcomes around a mean a bell curve.

It isn't just jelly bean samples that plot out this way. Many randomly taken data samples plot out forming a bell curve, although the steepness of the sides of the bell shape vary, depending upon the amount of deviation in the samples.

In statistics, by definition, 65 percent of the samples fall within what is called *one standard deviation from the mean* (or average), 95 percent of the samples fall within two standard deviations, and 99.7 percent of the samples fall within three standard deviations. The formula for determining one standard deviation is:

$$SD = \sqrt{\frac{\text{sum of (Entries - A)}^2}{\text{No. of Entries}}}$$

Where:

SD = Standard deviation

A = Average of the entries

Entries = Samples or data

Two standard deviations equal twice the standard deviation and three standard deviations equal three times the standard deviation.

To calculate the standard deviation for our red jelly beans we plug the variables into the standard deviation formula and yield a standard deviation of 2.47. This means that if we were to perform our sampling experiment again, we can expect that 65 percent of the outcomes will fall between 9.53 and 14.47 red jelly beans. Recalculating to the nearest whole jelly bean yields a probability that 77 percent of the outcomes will result in 9 to 15 red jelly beans being drawn, and 96 percent of the outcomes result in 7 to 17 red jelly beans. Finally, 99.5 percent of the time the number of red jelly beans will be between 5 and 19.

De Moivre's method allows us to calculate whether an outcome falls within a statistically meaningful range around an average. To put it another way, it allows us to determine mathematically which outcomes are more likely than others are.

Data that fall beyond the deviation range are called *outliers.* For example, if we are interested in all data within 95 percent of the average or two standard deviations, data that lies beyond 95 percent are outliers. While outliers beyond two standard deviations are possible, they are not very probable, their probability being less than 5 percent, or 2.5 percent on either side of the two standard distribution ranges. Therefore, if someone were to say that they pulled 20 jelly beans from our jelly bean jar and all 20 were red, we would have great cause to be skeptical, because this result would be well outside of three standard deviations, meaning it is very unlikely.

When things operate within specified ranges the outcomes are generally predictable and relatively free of risk. Operate outside of proven parameters and risk increases. The temperature light on our automobile dashboard lights up when the engine temperature exceeds a specified safe operating temperature. It warns us that something is wrong. If we ignore the light and continue to drive, we do so with added risk.

The same is true when design professionals or firms operate outside the normal ranges for the profession. For example, when a design firm spends more money than its competitors do trying to get work, makes more errors on its drawings, or has more projects over budget than other design firms doing similar work do, then the firm is adding risk. If nothing is done to bring the firm's performance within the range of acceptable standards, sooner or later, something bad is likely to happen. To assist design firms in knowing how they compare to others in the design profession, Chapter 9 includes some rules of thumb regarding performance characteristics of design firms.

The *Challenger* space shuttle tragedy is an extreme example of what can happen when an organization decides to operate outside of proven ranges of safe operation. The morning of January 28, 1986, was unusually cold for Florida. Temperatures during the night had fallen below freezing. On past launches, the company that designed and built the shuttle's solid rocket boosters (SRBs), Morton Thiokol, had experienced difficulty with the SRB o-rings sealing properly when temperatures were too cold. If o-rings did not seal properly, burn-through in the SRB joints was possible with potentially catastrophic results. Morton Thiokol engineers had determined that at temperatures below 53°F there was no certainty that the o-rings would seal properly. Morton Thiokol's engineers recommended postponement of the flight until the temperature warmed to within the calculated range of safe operation, but Morton Thiokol management overrode the recommendation. The rocket was launched with seven crew members aboard the orbiter. The o-rings in the right SRB did not seal properly. Seventy-three seconds after liftoff, flames burned through an o-ring joint causing structural failure of an external fuel tank. Contrary to what everyone thought he or she saw on television, there was no explosion. Instead, aerodynamic forces tore the rocket to pieces and the orbiter, still intact, plunged into the sea.

FACTS AND UTILITY ARE NOT THE SAME

"Facts are facts"—we hear this all the time, as if facts are all we need to know in order to make a decision. However, we do not make decisions based solely on facts, as Pascal's Wager illustrates. We make decisions based on what the facts mean to us, and their meanings are as varied as we are.

The 18th century is known as the Age of Enlightenment. During this period, reason reigned supreme. Eighteenth-century intellectuals of all stripes believed that reason was the key to objective truth. It was considered the basis for decision making in science, art, ethics, government, logic, and even in risk management.

Daniel Bernoulli (1700–1782), Jacob Bernoulli's nephew, is best known for his work in fluid mechanics and the principle that bares his name: For every increase in the velocity of a fluid, there is a simultaneous decrease in pressure. Besides fluid mechanics, Bernoulli had an interest in risk management, and in this subject Bernoulli swam against the tide. In 1738, he wrote a paper entitled "Exposition of a New Theory on the Measurement of Risk." In it, Bernoulli explained that ever since mathematicians began studying probability, they computed the values of risk by

> multiplying each possible gain by the number of ways in which it can occur, and then dividing the sum of these products by the total number of cases.[22]

Risk management was a straightforward math problem. As we have seen, it is the approach that Cardano used two centuries earlier and so did Bernoulli's uncle Jacob. By the 18th century it was taken for granted that risk management was done by the numbers. Bernoulli argued that this strictly mathematical approach failed to take into account the way normal people really make decisions. People take more than facts into consideration when making decisions about an uncertain future. According to

Bernoulli, utility plays a vital role in decision making. Utility is the usefulness of a decision based upon subjective factors that vary from person to person. Will a decision bring greater happiness? Will it bring greater wealth? Will it bring fame or improve my standing in the eyes of family, friends, and my professional peers? In affect, what is the decision's value to me? These factors cannot be quantified like facts. Nevertheless, they are important in decision making. There are facts and there are perceived facts and these have different meanings to different people.

For example, here is a fact: an eight-ounce glass contains four ounces of water. This fact alone, however, cannot lead to a decision or action. However, through the addition of utility—what the eight-ounce glass with four ounces of water means to me—a decision can be made. I can decide if it is half empty or half full, depending upon its subjective meaning to me, which can vary depending upon my circumstances and state of mind. Once I give it meaning, I can decide what to do with it. I can take a big gulp of the water because I believe I have plenty or I can sip it slowly to make it last because I don't have much water to spare.

Utility, Bernoulli claimed, drives decision making in the real world. He likened utility to a weight factor in a mathematical formula, a factor that varies among individuals and their circumstances. Bernoulli thought that personal wealth was a considerable factor in decision making and formulated a risk management principle based on it:

> The utility resulting from any small increase in wealth will be inversely proportionate to the quantity of goods previously possessed.[(23)]

We see Bernoulli's risk management principle at work nearly every day. On television game shows, contestants who have won very little are usually willing to risk it all and continue to play in hopes of winning even more. Once a contestant has won a great sum, however, he or she is usually much less willing to risk it all and continue playing. The principle is not specific, however; different contestants are willing to risk different amounts. Ten thousand dollars might be a lot of money to one, whereas another may be very willing to risk the $10,000 in order to make more.

For design professionals, Bernoulli's risk management principle explains why small design firms and large design firms have different appetites for risk. Small firms with little wealth have much to win. In contrast, large firms with considerable wealth have much to lose. This affects the way they make decisions about managing risk.

THE VITAL FEW AND THE TRIVIAL MANY

Vilfredo Pareto (1848–1923) was an Italian mathematician and economist. Pareto made many contributions to economics, but for risk managers, his significance comes from an observation he made about property in Italy. He noticed from census data and research that 80 percent of the land in Italy was owned by just 20 percent of the people. Pareto noticed that other things seemed to follow this 80/20 pattern. Pareto planted peas in his garden and observed that 80 percent of the big pea pods he picked came from just 20 percent of his pea plants.

Pareto's observation was mostly ignored until the 1940s when the American economist Dr. Joseph Juran stumbled upon Pareto's work. Building upon Pareto, Juran noted that 20 percent of manufacturing defects cause 80 percent of the problems. Juran called this phenomenon the "vital few and trivial many" principle or the 80/20 rule. It is also known as Pareto's law or Pareto's principle.

The 80/20 rule contains a valuable lesson for design professionals. It suggests that 20 percent of a design firm's projects will cause 80 percent of the firm's losses. Used another way, it suggests that 80 percent of the firm's profit will come from 20 percent of its projects. Design firm principals will spend 80 percent of their time working on problems associated with 20 percent of the projects.

The 80/20 rule suggests a method for effective risk management. Identify the handful of problems that pose the biggest threat and make sure these are managed properly because they account for most of the risk.

MONTE CARLO OR "WHAT-IF" MODELING

The Polish mathematician and physicist Stanislaw Ulam liked to play solitaire. In 1946, while playing solitaire and recovering from an illness, he wondered what was the probability of winning at the game. What if he could figure it out without having to play hundreds of times? Ulam hit upon an idea.

Electronic computers were new inventions at the time, but Ulam had access to a very powerful one because he was working with Edward Teller designing the hydrogen bomb. What if he used the computer to model the game of solitaire and then had the computer play solitaire over and over and thereby determine the probability of winning? Ulam shared his idea with two friends, John von Neuman and Nicholas Metropolis, and they began working on a computer program that could shuffle cards and play solitaire. Their program randomly generated numbers, which simulated the shuffling of cards. The program then played solitaire with the electronic equivalent of a deck of cards. In 1949, the three published an article explaining their random number generating process. They called it the *Monte Carlo method,* named after the casinos in Monaco. The Monte Carlo method is "what-if" modeling. It uses a computer to explore various "what-if" scenarios. By changing portions of the data, a variety of alternative hypothetical outcomes can be explored.

Months before Hurricane Katrina struck New Orleans in August 2005, a computer model predicted the devastation. In 2004, the Center for the Study of Public Health Impacts of

Hurricanes at Louisiana State University (LSU) ran a computer simulation to determine what might happen if a hurricane struck New Orleans. The LSU scientists called their "what-if" computer simulation Hurricane Pam. Pam was a Category 3 hurricane built with weather data provided by the National Weather Service. The computer model showed Hurricane Pam overtopping levees and flooding New Orleans. The computer model showed that the area around the intersection of the Intracoastal Water Way and the Industrial Canal near the Lower Ninth Ward was particularly vulnerable. When Hurricane Katrina slammed into New Orleans months later, levees in this precise area breeched, flooding both the Lower Ninth Ward and Orleans Parish. LSU made their computer simulation results available to the federal government and FEMA several months before Katrina struck.

Although they rarely use sophisticated computers to model alternative outcomes, design professionals use "what-if" scenarios all the time: "What if we chase this project—can we win? What relevant data do we need in order to determine our chances? What if we delete this column and increase the span of this beam? How deep will the beam be? What will the savings be in the cost of construction? What is the tradeoff in safety if we reduce redundancy in the structural system?" With spreadsheets, thought experiments, and pencil, paper, and calculations, design professionals explore various "what-if" scenarios and make decisions about design, construction, and their design firms.

PRAGMATIC INFERENCE AND RISK MANAGEMENT

The American pragmatic philosopher Charles Sanders Peirce (1839–1914) suggested that decision makers do not make decisions based on mathematical or scientific exactness. Instead, he argued that decision makers, risk managers included, make inferences based on a combination of experience and educated guesses.

Risk managers base their decisions about the future on past events assuming that the same rules of physics, logic, business, and, in the case of design professionals, professional practice will be more or less the same in the future. Risk managers infer that past rules and experience apply to the future.

There are three types of inference: induction, deduction, and what Peirce called *abduction.* While risk managers would readily admit that they use the first two, they may not be aware of the third at work in their decision making. Nevertheless, abduction is the most important of the three.

Deduction is the tool of inference in which the conclusion necessarily follows from one or more premises. Induction is the tool of inference in which a number of specific observations leads to a general conclusion regarding the observations.

Abduction is the tool of inference that yields an explanatory hypothesis based on observations or facts. It is educated guessing. Peirce explained it this way:

> A hypothesis...has to be adopted, which is likely in itself, and renders the facts likely. This step of adopting a hypothesis as being suggested by the facts, is what I call abduction. I reckon it as a form of inference.(24)

> If you carefully consider the question of pragmatism you will see that it is nothing else than the question of the logic of abduction.(25)

Although hypotheses are used by scientists, they are not reached scientifically. They are derived through abduction—educated guesses of what are likely to be the cases. Scientists then set out to verify or disprove their hypotheses. Like scientists, design professionals employ this method in managing their projects and practices.

Risk managers are pragmatic decision makers. Relying on personal past experience, experiences of the firm, and experiences from other firms, they do what they think is best to control some future event. The value of their decision is also at risk until the future reveals whether the idea worked. The value of an idea is determined by the outcome. Good ideas are only recognized later to be good, not when they are implemented. The greater the wealth of past experiences, the greater the likelihood that events will prove that the ideas used to control risks were good ones.

CHAPTER SUMMARY

In this chapter we discussed the fundamental principles and intellectual tools used in risk management:

- The use of fractions to express probabilities: Some outcomes are more likely than others are. Risk managers implement procedures that make the more desirable outcomes the more likely outcomes.
- Risks and rewards are not always proportional. Generally small risks yield small rewards and big risks yield big rewards, if successful. However, there are also opportunities for big rewards with little risk.
- The whole can be estimated from the parts. The performance and characteristics of other design firms can be used as guideposts to measure your firm's performance.
- Facts and utility are not the same. Design professionals do not make decisions based solely on facts. Instead, design professionals make decisions based on what the facts subjectively mean to them.
- Things are predictable within a range. If your design firm's performance or characteristics are far outside of the normal range of your competitors, then determine why. There may be grave risk in continuing to operate far outside the norm.

- Managing the vital few takes precedence over managing the trivial many. Determine the handful of risks that pose the greater threat and concentrate on managing them.
- Abduction is the pragmatic inference tool of logic. The other two logic inference tools are induction and deduction. Abductive decisions are founded on educated guessing based on the review of available facts and data. Because all relevant data are rarely available when a decision needs to be made, abduction is the fundamental and most useful tool of risk management.

Chapter

2

THE RISK MANAGEMENT PROCESS

The playwright William Stoppard began his 1967 play *Rosencrantz and Guildenstern Are Dead* with Guildenstern taking a coin from his money bag and tossing it. His friend, Rosencrantz calls out "heads," observes that the coin lands heads, and then puts it in his own money pouch. The audience watches as the two repeat the process over and over. Eventually Rosencrantz announces that the coin has landed heads 85 times in a row, claiming that this has "beaten the record!" Reflecting upon his phenomenal run of bad luck, Guildenstern asks Rosencrantz what he would think if the coin had landed tails 85 times in a row. Rosencrantz responds, "Well, I'd have a good look at your coins for a start!"(26)

This scene contains two important messages for risk managers. First, Guildenstern's bad luck is Rosencrantz's good fortune. Every coin has two sides. When a coin lands heads up, it simultaneously and—just as importantly—lands tails down. Good fortune and bad luck are two sides of the same coin. The second message connects to the answer to the following questions: "Is it possible to control how the coin lands? That is, can we influence the probability of outcomes?" Rosencrantz's answer suggests that the only way of doing so is to cheat. Yet Stoppard's play, which is based on two minor characters, Rosencrantz and Guildenstern, in Shakespeare's play *Hamlet,* suggests another possibility: Ultimately the world might be fated, outcomes predetermined, and all we can do is find our own personal meaning in it. In Stoppard's play, Rosencrantz and Guildenstern do a very bad job of finding meaning in their world. In scene after scene they try in vain to understand what is going on around them. Often they sense that somehow they are being manipulated, but fail to take steps to exercise any control. Just before he finally disappears from the stage (vanishes from the world) Guildenstern remarks, "There must have been a moment, at the beginning, where we could have said no. But somehow we missed it."[(27)] This suggests that these men could have—should have—done something to control their fate. Had Guildenstern been a risk manager, he would have known with certainty that there was such a moment—indeed many moments—when he and Rosencrantz could have done something to control their fate.

Risk managers don't cheat, but they do manipulate the odds in their favor in two ways. They reduce the probability of bad things happening and they increase the likelihood that good things will happen. Combined, these two opposites control the way the coin is tossed and thereby maneuver outcomes to the favor of the risk manager. Right actions reduce risk.

The Auditorium Building example in Chapter 1 illustrated only one aspect of risk management for design professionals, construction-related risk management. There are many other risks that affect the modern day practice of architecture, engineering, and planning. Today, the design professions must function as well-rounded businesses. For a design firm to succeed over an extended period of time, it must successfully manage the many risks associated with all facets of the profession and business of design. This includes risks associated with project management; design; production of construction documents; marketing design services; hiring, retaining, and providing career opportunities for qualified staff; client and consultant agreements; professional liability insurance; design fees; project and construction cost control; project schedule control; and providing construction support services.

THE STEPS OF RISK MANAGEMENT

In Chapter 1, we looked at the tools and principles of risk management decision making. Now we apply them to the risk management process. The process is methodical, procedural, and repeatable. It is based on an understanding of probability, logic, and pragmatic decision making regarding unknown future events. The process can be improved over time as experience grows and data improves. It can be easily learned and implemented by anyone interested in exercising practical control over future possible outcomes.

There are six fundamental steps in the risk management process:

1. *Acknowledge risks:* Recognize that there are many risks that affect the practice and business of design and that these risks can be controlled by a comprehensive risk management plan.
2. *Determine the risks:* There are many risks that adversely affect success. They can be grouped into 15 risk-source categories.
3. *Determine the probability:* The chances of various risks occurring vary depending upon particular circumstances. Probabilities can be determined quantitatively or qualitatively.
4. *Determine the severity of the impact:* Quantitatively or qualitatively predict the consequences if certain threats materialize.

5. *Develop procedures and solutions to manage risks:* Follow the six basic strategies used to manage risk.
6. Implement those solutions or procedures that manage the risks that have the highest overall negative impact on the desired outcome. Develop metrics for measuring results and monitor the effectiveness of the implemented solutions and procedures.

Step 1: Acknowledge Risks

The first step in the risk management process is to recognize that something can be done to manage future outcomes—good and bad. Although this sounds simple, Rosencrantz and Guildenstern never arrived at this conclusion. The "che sarà sarà" syndrome plagues many people: "what will be, will be." Even worse is the mantra: "There is no sense worrying—nothing will turn out okay." For some, this first step in risk management is difficult. It is tantamount to a leap of faith. Nevertheless, design professionals who want to manage risk successfully must take this initial first step. The success of what follows depends on it. A half-hearted attempt to manage risk is likely to fail. It will not be comprehensive or consistent. It is likely to be reactive, always putting fires out rather than preventing them from happening in the first place. Design professionals must embrace the idea that risks can be controlled and managed, and then proceed to develop and implement plans that do just that.

Step 2: Determine the Risks

The second step is to identify all of the risks that adversely affect a successful design practice. There are many and they are associated with all facets of the practice of design: A project can go over budget; a key employee can quit just before an important deadline; an improper flashing detail can cause a roof to leak; and so on.

To manage risks, it is necessary to know what they are and from where they are coming. One way to do this is to make a list of all the possible things that can go wrong. In a sense, write down a grocery list of risks. The problem with the grocery list approach is that the list will go on for several pages without any particular order. It will be unwieldy, making it nearly impossible to figure out a meaningful way to manage all the unrelated risks. It is much better to organize risks by their sources. Arranged in this manner, it is easier to develop methods for managing them. The following list groups risks into 15 general risk-source categories that are common to the practice of design.

1. *Project management:* These are the risks associated with the process of managing design projects, including their scopes, budgets, and schedules. This risk-source category includes all risks associated with managing design teams. It also includes risks associated with the character, abilities, and shortcomings of project managers.
2. *Technical performance:* These are the risks related to project performance including project programming, planning, design and the production of contract drawings, specifications, cost estimates, and reports and other technical services and deliverables. This also includes the risks associated with the transition from iterative design to linear production and the risks connected to code and zoning noncompliance.
3. *Organizational:* These are the risks associated with the way the firm is organized and managed, including risks related to office procedures and politics.
4. *Business development:* These risks relate to marketing design services, producing proposals, statements of qualifications, brochures, and the costs of doing business development.

5. *Financial:* These are risks resulting from the firm's overhead rate, utilization rate, aging accounts receivable, profit, borrowing to meet payrolls, investments, and so on. Risks associated with project budget and financial performance are included in the project management category as well.
6. *Contractual:* These are the risks associated with the language of contracts, types of fees, payment provisions, retainers, indemnification clauses, and other agreement requirements.
7. *Insurance:* These are risks associated with carrying or not carrying professional liability insurance (sometimes called errors and omissions or E&O insurance). This category includes risks associated with claims against the design firm.
8. *Quality:* Risks associated with the mistakes, omissions, and the general lack of quality associated with work products are quality risks.
9. *Employee:* These risks pertain to the hiring, training, retaining, promoting, and professional development of staff.
10. *Consultants:* Risks associated with the selection and uses of consultants to provide services or perform certain portions of the work are considered consultant risks.
11. *Client and project types:* These are the risks that pertain to doing work for specific or general types of clients and projects.
12. *Technological and work product standards:* These risks are associated with the hardware and software used to produce the work. They include computer drafting standards, specification, and cost-estimating standards and procedures or the lack of them.
13. *Bid and award:* These are the risks that result from participating in pre-bid meetings, responding to contractor's questions during the bid period, preparing addenda, bid openings, evaluating bid results, and making recommendations to owners regarding awarding construction contracts.

14. *Construction phase:* These are risks related to the review of shop drawings, responding to contractors' requests for information (RFIs), reviewing contractors' requests for pay, construction visits, project closeout, and working with owner-hired construction management organizations.
15. *External:* These risks are beyond the control of the design firm such as risks associated with the upturns or downturns in local, state, or the national economy. It includes risks associated with changes in technology, code standards, changes in the market areas, government policies, and so on.

The risk-source categories listed are not in order of importance. In fact, their importance and impact varies, depending upon the nature of the design practice, its size, geographic location, types of projects, and a host of other factors. Knowing the sources of risk, however, is an important step in any comprehensive risk management plan.

Step 3: Determine the Probability

The third step is to determine the likelihood of a threat, or a risk, occurring. Insurance companies use probabilities to manage their exposure to financial risks. An insurance company might look at historical records and calculate that out of every 10,000 insurance policies for a particular type of coverage, it averages 700 claims per year, yielding a probability of a claim occurrence of 700/10,000 or 7 percent. Because the insurance company has a lot of data to work with, it can also determine the likely variance of future claims from this mean of 7 percent, by calculating the standard deviation. As we learned in Chapter 1, the insurance company can be "exactly as possible certain" (using Jacob Bernoulli's term) that all future claims will fall within three standard deviations from this mean of 7 percent. This is called a *quantitative analysis,* meaning it uses hard, objective, historical data to ascertain exact probabilities that are within statistical, quantifiable ranges. Quantitative analysis relies on numbers and number crunching to derive precise answers.

For design professionals, however, hard, objective data is not always available. Many design-related risks cannot be quantified with statistics. For example, the risk of going after a particular project and failing is usually not exactly quantifiable. If it were, design firms would know in advance with "moral certainty" (another Jacob Bernoulli term) which projects they can win and which they cannot. Consequently, they would only go after those projects they know they will win and not bother chasing those they know they will not. The design professions would be far less risky businesses if this were the case. However, the design industry does not work this way. It is possible, though, to use qualitative analysis to improve the chances of winning, while simultaneously decreasing the chances of losing. Qualitative analysis is determining probabilities by subjective judgments. The probabilities are not as exact as quantitative ones. They are not based on objective data. Instead, probabilities are dependent on subjective conclusions about past events and experiences. Qualitative judgments about probabilities are far more common in the design professions than quantitative ones. Most of the risk management suggestions in the latter chapters of this book are based on qualitative judgments about risk—judgments based on anecdotal evidence from personal experiences and the experiences of others, not on hard statistical evidence. Later in this chapter, we take a look at a few common qualitative approaches for determining probabilities.

Step 4: Determine the Severity of the Impact

The probability of something bad happening is only half of the risk-exposure equation. The other half is its impact. Like probability, impact can be measured quantitatively or qualitatively. In the previous example, the insurance company's historical data showed that it averaged 700 claims per year for every 10,000 insurance policies. Suppose other insurance company data showed that the claims averaged $2,500 each. This $2,500 is the impact and it represents the company's financial risk exposure per claim. A little arithmetic yields a total financial risk exposure of $2,500 × 700 claims = $1.75 million per year for every 10,000 insurance policies. This is an example of a quantitative analysis.

Qualitative analysis does not rely on historical data; it relies on judgments based on past experience, as discussed. We take a look at ways to do this later in this chapter.

Step 5: Develop Procedures and Solutions to Manage Risks

Now that risks have been identified and their probabilities and potential impacts determined, the next step is to develop procedures or solutions that reduce, eliminate, or manage them. There are two approaches: We can be proactive and figure out beforehand how to deal with the risks or we can be reactive and deal with their consequences. The former manages the causes of risk, and the latter responds to their effects and deals with the fallout. Cleaning up after something dreadful happens is often more difficult and costly than preventing it from happening in the first place. As we know, it is difficult to put toothpaste back in the tube after it has been squeezed out. Proactive risk management is founded on the aphorism that an ounce of prevention is worth a pound of cure.

There are five fundamental strategies for the proactive management of risk and one for managing the aftermath or effects. The proactive strategies are: prevention, mitigation, transference, contingency planning, and assumption. The reactive strategy is recovery planning.

1. Prevention—reduces risk by reducing the number of uncertainties and/or threats.
2. Mitigation—lessens the impact of the uncertainty or threat should it occur.
3. Transference—passes or transfers the risk to some other party, making it responsible for the consequences of the risk.
4. Contingency planning—takes a proactive approach that requires the advanced preparation of a plan for dealing with specific uncertainties or threats should they occur.
5. Assumption—requires an acceptance of the potential impact of the risk because the cost of preventing, mitigating, transferring, or developing contingency plans for dealing with it is greater than the possible negative effects.
6. Recovery planning—uses reactive planning that deals with the effects or outcome of a risk management strategy gone wrong. When something unexpected happens that causes a threat or a great increase in risk, something must be done to rectify the situation.

Depending upon the risk, one or more of these strategies can be used. Prevention or mitigation can best be accomplished by developing a quality assurance plan and applying it to all work in the office. For example, there is risk associated with design firms directly hiring geotechnical consultants and surveyors. Transference is a strategy for managing this risk. Many design firms suggest that their clients hire the geotechnical consultant and surveyor directly, thereby transferring the risk to the client, although not all clients will accept the risk.

An example of contingency planning is to budget a sum of money for the replacement of water-damaged (dry-rotted) materials for wood-framed building remodeling projects. Even if there are no signs of dry rot from looking at the building's finish materials, wood-framed buildings may have concealed areas of dry-rotted plywood, studs, or joists that are not apparent until exposed during construction. Replacing dry-rotted materials can be costly and if not budgeted for, can adversely affect the project's construction budget. Although the presence of dry rot is often uncertain, it is foolhard not to have a contingency to cover the costs of replacing dry-rotted materials because the threat is so common.

Some risks are small and therefore better just to live with should they occur. For example, the client may request an additional meeting. Some clients are "meeting freaks" as too many design professionals have found out the hard way. Meetings can be expensive in both labor time and travel costs. Should an additional meeting be requested, the design firm may find, under certain circumstances, that it is in the firm's best interest to just assume the cost of the additional meeting in exchange for the greater benefit of client goodwill. Some scopes of services stipulate the number of client meetings that the design professional shall attend in order to reduce the risk of design cost overruns due to too many meetings.

An example of contingency planning is setting aside a budget contingency for design work before the work ever begins. No project manager is perfect and unexpected things always happen during the design or production of construction documents. Because no one can foresee the future with certainty, a contingency fund is set aside and the work planned to finish without using it.

Recovery planning is responding after something had gone wrong. The Chicago architect Daniel Burnham was in charge of overseeing the design and construction of the fairgrounds for Chicago's Columbian Exhibition of 1893. Because dozens of buildings had to be designed, it was more work than one architectural firm could handle so Burnham prepared a list of qualified architectural firms to assist. Although Burnham's design firm, Burnham and Root, was from Chicago, Burnham felt that Chicago's architectural firms (today we call it the Chicago School) were not on par with the architecture firms of Europe

FIGURE 2.1 *Transportation Building, World's Columbian Exposition. Adler and Sullivan, architects. Built 1891-1893. View 1893 R. Cates, photographer. Historic Architecture and Landscape Image Collection, Ryerson and Burnham Archives, The Art Institute of Chicago. Digital file © The Art Institute of Chicago.*

and the East Coast of the United States. At the time, Beaux Arts Classicism dominated the world of architecture. Burnham was a self-taught architect, having never attended architectural school. Perhaps for this reason, he felt that what Chicago architects were doing—experimenting with tall glass and steel commercial buildings—was not appropriate for a world exhibition. Consequently, he turned to East Coast firms for assistance.

When Burnham's architect peers in Chicago found out that Burnham was inviting East Coast offices to design buildings in Chicago, they were considerably upset and offended. Louis Sullivan was furious. Burnham quickly realized his mistake and instigated a recovery plan. He revised his list to include Chicago architects. Chicago's architects accepted Burnham's revised plan, although Sullivan was reluctant to do so at first. Finally, however, he, too, agreed to participate and eventually became the architectural steering committee's secretary. Burnham's recovery plan resolved the threat. By most accounts, the only fairground building of any architectural repute was the Transportation Building designed by Adler and Sullivan (see Figure 2.1).

Step 6: Implement Solutions/Procedures to Manage the Greatest Risks

Steps 1 through 5 constitute the development of a risk management plan. The last step involves implementation of the risk management plan. To work, a plan must be implemented. In addition, its results must be monitored to determine how it is progressing and if things are going as expected. Depending upon the monitored results, fine tuning or modifications to the plan may be necessary. This all requires commitment and constant attention. It requires all firm management members to support the process and it requires participation from everyone in the office. Suppose, for example, that a design firm decides no work shall be performed without a written agreement. For design firms that work frequently with new clients, this may be a very reasonable risk management procedure. Perhaps the firm has been burned too many times before, starting work before an agreement is signed, only to find out later that the client changes its mind and doesn't sign it. This is not uncommon. A variation of this is to get too far ahead in a project. Some projects require clients to initiate notices to proceed at various milestones. For example, a project may require the client to issue a notice to proceed after schematic design. If the client does not issue the notice to proceed, but the design firm continues working, getting paid for the work is at risk. The client may choose to drop the project for some reason and not issue the notice to proceed for the next phase. If it does, the design firm has wasted its time and money. Therefore, the firm institutes a policy: No work without an agreement or notice to proceed. However, suppose the procedure is routinely ignored. If no one is monitoring what is happening, the firm may be unaware of the extent to which this risk management policy is being violated. Although the firm has the policy, it is not actually implemented and consequently it is not managing the risk. Implementation, commitment, and monitoring are crucial to the success of a risk management plan.

DETERMINING OVERALL RISK

In the previous insurance company example, we multiplied the probability of an insurance claim (7 percent) by the average impact of a claim ($2,500) to determine the company's yearly exposure to claims ($1.75 million). This exposure is called *overall risk.* Because the insurance company had hard, objective data, it was possible to determine the overall risk quantitatively. Yet sometimes the overall risk must be determined qualitatively—in practice, qualitative analysis is more common. The following example illustrates an approach for qualitatively determining probability, impact, and overall risk.

The probability of a concrete slab forming thin cracks when it cures is relatively high. Concrete shrinks when it cures and this causes hairline cracks to form. The exact quantitative probability of this happening is unknown, but we can calculate the qualitative probability. More than one structural engineer has said that there are really only two types of concrete slabs: those that have cracked already and those that are going to. We'll pick a probability of 99 percent for a slab developing hairline shrinkage cracks, because there is always the slight chance that it won't.

The next step is to determine the impact of hairline cracks. What are the negative consequences should they occur? In general, hairline cracks caused by curing are cosmetic. They affect a concrete slab's appearance and have no real impact on its structural integrity. If the plan is to cover the floor with carpeting, no one will ever see the cracks if they occur. Consequently they may have no negative effect, whatsoever. In this case the impact may be 0 percent. However, suppose the plan is to leave the concrete exposed, apply a colored stain to it, and use the slab as the finished floor. In this case, the stain will magnify the cracks because the cracks will stain darker than the surrounding concrete floor. A lot of randomly occurring stained cracks may be unsightly and undesirable. It may even make some people think that something is structurally wrong with the slab. Assuming that random and unsightly cracks are undesirable, let's apply a higher risk impact percentage to the cracks, say 80 percent.

Overall risk is determined by multiplying the probability of the risk by its impact. If we are going to cover the concrete slab with carpet we determine that the overall risk is 99% (0.99) × 0% = 0%. However, if we are going to leave the slab exposed and stain it, the overall risk might be 99% (0.99) × 80% (0.80) = 0.79%. The product 79 percent is called the *overall risk factor*.

Although the probability of the cracks occurring is the same, their impact varies depending upon our judgment of the situation and, consequently, this affects the magnitude of their overall risk. Using Daniel Bernoulli's term, overall risk is a measure of utility; it is a measure of its subjective meaning.

The next step is to determine what, if anything, can be done to manage the overall risk. We could ignore it and do nothing—in affect, assuming the risk. If carpeting is going over it, assumption is probably the best strategy, and our time is better spent worrying about risks with higher overall risk factors. However, in the case of exposed stained concrete the overall risk is high, 79 percent. Consequently, something should be done about it. Consider the other proactive risk management strategies: prevention, mitigation, transference, and contingency planning. We could employ one or more of them to manage the overall risk. The strategy that makes the most sense in this case is mitigation—lessen the impact of the risk should it occur. We have already determined that it is nearly certain to occur, so it makes the most sense to figure out a way to control where the cracks occur so that they do not adversely affect the appearance of the finished stained concrete floor. There are many design details that structural engineers employ to control where hairline shrinkage cracks occur. Through a combination of control joints and construction, or cold, joints it is possible to limit cracking to specific locations within these joints, making them virtually invisible to the average person. In this case, mitigation is the preferred strategy for controlling the 79 percent overall risk factor of hairline cracks in a stained concrete floor.

We have now concluded a risk assessment for just one possible project-related risk. There are many more project-related risks that require similar analyses. Those with high overall risk factors need to be managed using one or more of the risk management strategies. It is not just project-related risks that should be studied in this manner; all aspects of the practice and business of design include risks that should be analyzed for their overall risk potential.

It may not have been apparent, but there is a problem with the previous method used. We used 99 percent for the probability of the hairline cracks occurring because the likelihood was very high. We could have used 90 percent or 95 percent. In addition, 80 percent was arbitrarily chosen for the impact of cracks for stained concrete floors. Because there was no objective data to go by, only past experience, whatever percentage we chose would be a bit arbitrary. Nevertheless, it is unlikely that anybody on the design team would have quibbled much over 90 percent, 95 percent, or 99 percent for the probability of cracks occurring. However, there may have been a few eyebrows or even some argument about an 80 percent assignment for its risk impact.

The problem with assigning specific but arbitrary percentages to risk probabilities and impacts is threefold. First, the specific percentages imply more accuracy than there really is. Second, assigning specific percentages can lead to needless and often protracted debates among team members—"should it be 75 percent or 85 percent?" Rather than solving problems, the team winds up debating percentages. Third, there are many risks associated

RISK ASSESSMENT TABLE

Description of Risk	Probability (3-category)	Impact (5-category)	Overall Risk Factor
Moisture Incursion (high water table)	3	3	12
Differential Settlement (poor soil conditions)	9	6	54
Cracks in Slab on Grade	9	1	9

FIGURE 2.2 *Typical risk assessment table.*

with all projects and when the percentages used to determine all of them are arbitrary, the design team cannot know with confidence what risks really pose the greater overall threats, especially when their overall risk factors are close to each other. To solve this problem, there is another approach that is more suitable to qualitative risk analysis.

Instead of specific percentages we can assign probabilities three distinctive numbers: 1, 3, and 9, where:

- 1 = low probability, 10 percent or less chance of occurring
- 3 = medium probability, 20–50 percent
- 9 = high probability, greater than 50 percent

Generally speaking, it is not difficult for an individual to determine, or for a team to agree, whether a risk has a low, medium, or high probability of occurring. Once that determination is made, its numerical value is pre-assigned 1, 3, or 9. In the

IMPACT

PROBABILITY	1	2	4	6	8
9	**9**	**18**	**36**	**54**	**72**
3	**3**	**6**	**12**	**18**	**24**
1	**1**	**2**	**4**	**6**	**8**

Highest 20%

Lowest 20%

FIGURE 2.3 *Risk assessment table showing the 15 possible overall risk factor ratings. Overall risk factors 1, 2, and 3 represent the lowest 20 percent and overall risk factors 36, 54, and 72 represent the highest 20 percent.*

case of hairline cracks, this risk clearly has a high probability, so the numerical value for its occurrence is 9.

Note that the numerical values are geometric. As such there will be considerable gradation in the overall risk factors that will result, making those that pose the greater risk pop out like sore thumbs.

If three categories seems too crude and you would like a little more refinement in the analysis, use more categories. Remember though, this is qualitative risk analysis, not rocket science. The point is to have enough information to make decisions about risks, but not to get bogged down in the process of determining which risks are slightly more worth managing than others are. All you want to do is decide which to manage, not which ones are a few percent more risky than others are. The whole process is qualitative—the numbers do not mean anything in and of themselves. They are only helpful if they assist in making risk management decisions. Moreover, if the process is too refined

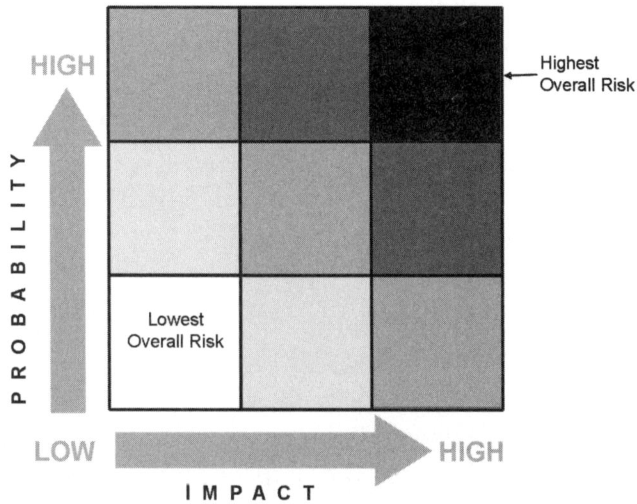

FIGURE 2.4 *Overall risk assessment matrix. The matrix assigns risks to squares within the matrix. Those of lowest risk reside in the lower left corner and those of highest risk in the upper right corner.*

or too complicated and/or time consuming, it will not be used. Nevertheless, five categories is another commonly used system. Rather than a geometric progression of 1, 3, 9, 27, 81, though, a much simpler progression makes the multiplication easier: 1, 2, 4, 6, 8, where:

- 1 = very low probability, 10 percent or less chance of occurring
- 2 = low probability, 10–30 percent
- 4 = moderate probability, 30–60 percent
- 6 = high probability, 60–90 percent
- 8 = very high probability, 90 percent or above

The same approach can be taken for risk impacts. They can be assigned numbers, again 1, 3, and 9, where:

- 1 = small impact—the impact is an inconvenience or annoyance only
- 3 = medium impact—the impact is more than an inconvenience, annoyance, or problem, but its negative impact does not jeopardize the success of the project or undertaking
- 9 = big impact—the impact can threaten the overall success of the project or undertaking

Alternatively, a five-category system can be used, where:

- 1 = very small impact—the impact is inconsequential or nearly so
- 2 = small impact—the impact is an inconvenience or annoyance
- 4 = medium impact—the impact is more than an inconvenience, annoyance, or problem, but its negative impact does not jeopardize the success of the project or undertaking
- 6 = big impact—the impact poses a great threat to the project or undertaking
- 8 = very big impact—the impact threatens the overall success of the project or undertaking

Using the three- or five-category risk impact ranges, it is not difficult to decide on the level of impact for a particular risk. As with risk probability, once the determination is made, the numerical value is pre-assigned. The number designations are geometrical (three-category range) or arithmetical (five-category range) to provide gradation in the resulting overall risk factors.

In our hairline crack example, using the three-category range, we determined that the probability of cracks occurring was 9—a high probability. Using the three-category range, the impact of cracks is a 1 if covered with carpet and a 3 for the stained concrete floor scenario. Based on these probabilities and impacts, the overall risk factors are:

- Carpeting: 9 (probability) × 1 (impact) = 9
- Stained concrete: 9 (probability) × 3 (impact) = 27

This approach can be used to study any number of risks. The analysis can be done using a spreadsheet computer program. The spreadsheet allows for the easy addition of risks and for the sorting of them by probability, impact, and overall risk. The three-category range and the five-category range can also be used together as shown in Figure 2.2 on page 49. The figure shows a typical risk assessment for three common project-related risks.

Note that in Figure 2.2 the product of the three- and the five-category risk impact range is the overall risk.

After determining the overall risk factors, the next step in the risk management process is to decide which risks require management and what strategies to use to manage them. It is possible to reach the conclusion that all risks should be managed, but this may not be practical, or necessary. The overall risk factor of some risks is so low it may be better just to assume the risks. It may actually cost more in time, energy, and money to manage them than the risks warrant. How many and which risks you manage is up to you, however, the risk analysis itself suggests an answer.

When the three-category range and the five-category range are used together, an interesting result takes place, which is illustrated by Figure 2.3 on page 50.

Recall the 80/20 rule: if you manage the 20 percent of risks that are most critical, you will solve 80 percent of your problems. The 80/20 rule is an important principle used in risk management decision making. In his book *Making Decisions,* management consultant Robert Heller commented:

> Common sense suggests that some factors in a [risk management] process are more important than oth-

> ers—and analysis supports this. In reality, only 20 percent of activities may account for up to 80 percent of results. This is known as Pareto's law, the "80/20 rule"...Pareto's law concentrates on the significant 20 percent and gives the less important 80 percent lower priority.[28]

Applying the 80/20 rule suggests that managing the risks with overall risk factor ratings of 36, 54, and 72 might be the most effective thing to do. Managing the top 20 percent may very well result in successfully managing 80 percent of the problems.

Another approach and application of the 80/20 rule is to pick the 20 percent of risks with the highest overall risk rating. This is slightly different than managing the risks with overall ratings of 36, 54, and 72. In some instances, managing the top 20 percent of risks may include risks with lower ratings than 36. On other occasions, it may mean that some of the risks with overall risk factor ratings of 36 will not be managed. In extreme cases, it may mean not managing some or all of them with overall risk factor ratings of 54. Of the two approaches, I personally prefer the former—managing all the risks with overall risk factors of 36, 54, and 72 even if more than 20 percent of the risks have ratings that are this high.

There are variations to the risk assessment spreadsheet illustrated in Figures 2.2 and 2.3. There is another risk management method that does not use numbers at all. Instead it uses an overall risk assessment matrix that looks very similar to the table shown in Figure 2.3. The risks are arranged in the matrix based on their probability (low, medium, high) and their impact (low, medium, high). The risks are assigned a location in the matrix. Because of the format of the matrix, risks that have the lowest overall risk assessment occur in the lower left corner and those with the highest occur in the upper right corner (See Figure 2.4 on page 51.).

Using the matrix method, it is clear that those risks in the upper right corner should be managed. Like the spreadsheet method, however, it is still a judgment call as to what other risks deserve attention and should be managed.

As these discussions illustrate, there is a considerable amount of subjectivity involved in risk management. Risk management is pragmatic. It is the process of knowing which risks pose the greatest threats based on personal past experience and the experiences of others. Through trial and error, the design professional learns lessons from experiences and events that turned out right and from those that turned out wrong.

In the design professions, there is always something new to learn. When you learn it, it's added to your pragmatic data bank, and you will apply it when the circumstances are right. Risk management decisions are made this way every day. In the chapters that follow, there are dozens of suggestions about managing the various risks that affect the design professions. You can add them to your pragmatic data bank and use them as you see fit when it is time to make a risk management decision.

The following story took place long before the term "risk management" was used. The story's protagonist, Zachariah Allen, never used a risk management spreadsheet or matrix; he relied on his experience, judgment, and risk management attitude to make decisions about controlling risk. Today, many design professionals make risk management decisions in a similar way. In Allen's case, he was concerned about the risk of fire in his textile mill.

A Risk Management Story

In the early 1800s, the concept of loss prevention was virtually nonexistent among American fire insurance companies. Insurance companies used industry statistics to set premiums and calculate their expected losses. No thought went into preventing or reducing losses by managing the risk of fire—fires happened or they didn't. For different industries, insurance companies calculated the probability of fires and determined premiums accordingly.

By the 1820s, the United States was growing by leaps and bounds and the cotton industry was booming. Textile mills sprung up like weeds along waterways throughout New England. In general, the mills were two- to four-story tall wood-framed buildings with floor cutouts for waterwheel-powered machinery and pulleys. The exterior walls of some mills were brick or stone, nevertheless, the floors and roofs were framed with wood. Wood construction made the mills inexpensive to build and easy to modify to accommodate additions and changes in workflow, machinery, and pulley layouts.

Fires plagued these mills. With wood-framed construction and numerous floor cutouts, small fires often grew into raging infernos. Fire insurance premiums for textile mills soared in response to the high risk.

In 1822, Zachariah Allen (1795–1882) began designing and building a textile mill in Allendale, Rhode Island. Over the course of the next 17 years, he made many improvements and additions to it, finally completing it in 1839. The building measured approximately 40 feet wide by 160 feet long. It was four stories tall plus an attic. The exterior walls were stone. At the time, it was one of the largest mills of its kind.

Allen, an entrepreneur, inventor, and engineer, was also a pioneer in risk management. His idea was to reduce the probability of fire in his mill by building various fire safety features into it. By doing so, he reasoned he could exercise control over his losses and lower his fire insurance premiums.

Allen constructed a large, square stone tower, centering it along the front, long side of the mill. At the top of it, he built a cupola and installed a fire bell to warn employees in the event of a fire. At the base of the tower, he installed a large water tank with a rotary fire pump. Connected to the pump was a copper pipe fire hose system that extended throughout the mill, the first of its kind in a U.S. textile mill. The tower also included exit stairs that served all floors of the mill. At each floor level, access to the stairs was through steel fire doors, again a first. Having made all these fire safety improvements, Allen figured he was entitled to a cost break for his fire insurance premium. In 1835, he asked his insurance carrier for a fire insurance premium rate reduction—and was turned down. The insurance company had no way to evaluate and quantify Allen's fire risk-reduction improvements.

Disappointed, Allen contacted other textile mill owners and convinced many of them to join him in controlling their losses due to fire and the cost of their insurance premiums. Together, they started the Manufacturers Mutual Fire Insurance Company. The company's founding principle was that losses due to fire can be controlled by reducing the risk of fires through good design and the employment of other fire loss-prevention measures. The company insured only what it called good risk factories, meaning those factories that employed various loss-prevention methods. In the company's first 14 years, Manufacturers Mutual policyholders' premiums averaged about one-half as expensive as the cost of premiums of other insurance companies.[29]

Allen was a fastidious risk manager, who believed that risk management was a continuous, ongoing effort. Always interested in new ways of reducing risk, he employed yet another loss-prevention measure in his 1839 Allendale Mill addition—slow burning construction.

Slow burning construction means that when something catches on fire, the fire burns and spreads slowly. This allows more time for people to escape the building and it gives firefighters more time to extinguish the fire. Slow burning construction is not the same as fireproof construction. Fireproof construction uses materials such as masonry, steel, concrete, and cement plaster. Slow burning construction is wood construction, but the floor framing is done in a particular way.

Prior to the 1830s, typical mill floors were an assembly of wood beams, closely spaced wood joists, and two-inch-thick planks spanning over the joists. Fire made easy work of these assemblies because the floors were not thick and the spaced two-inch-thick joists had large surface areas. Allen eliminated the joists and, in their place, installed a solid floor of wood made of 2 × 4 boards standing on end, placed one right next to the other. Consequently, the floor was a solid four inches thick, spanning from beam to beam. With less surface area and thicker floors, the floor burned much slower.

The story about Allen and his textile mill illustrates a historic and practical application of risk management. Allen, the risk manager, made reasoned and calculated improvements to his mill to reduce the risk of loss of life and property due to fire. He employed the strategies of prevention and mitigation, and planned to transfer risk to his insurance carrier. When the implementation of this strategy did not yield the lower insurance premiums he desired, he employed yet another risk management strategy, recovery planning—although a rather bold and adventurous application of it. He started the Manufacturers Mutual Fire Insurance Company. Eventually, Manufactures Mutual changed its name to Factory Mutual. Today, it is called FM Global, with insurance offices throughout the world.

CHAPTER SUMMARY

Design firms can control their fate by managing risk. Through deliberate actions design firms can reduce the chances of undesirable events happening and increase the probability that desirable ones will. The risk management process has the following six steps.

1. Acknowledge that risk can be managed through deliberate actions
2. Identify risks that adversely affect success
3. Determine the probability of risks occurring
4. Determine the impact should risks occur

5. Develop solutions and procedures to manage risks
6. Implement solutions and procedures to manage those risks that pose the greatest threat and monitor their effectiveness

There are many risks that affect the design professions. They can be grouped into 15 risk-source categories:

1. Project management
2. Technical performance
3. Organizational
4. Business development
5. Financial
6. Contractual
7. Insurance
8. Quality
9. Employee
10. Consultant
11. Project type and client type
12. Technology and work product standards
13. Bid and award phase
14. Construction phase
15. External

The probability of a risk occurring can be determined quantitatively or qualitatively. Quantitative probabilities are objective, based on historical data and statistics. Qualitative probabilities are subjective and use abduction (reasoned and educated guessing based on experience). The impact of a risk can also be determined quantitatively or qualitatively. Of the two methods, the qualitative method is more applicable and used more frequently in the design professions.

There are six proactive strategies for managing risk and one reactive strategy:

1. Prevention
2. Mitigation
3. Transference
4. Contingency planning
5. Assumption
6. Recovery planning—reactive

Risk management planning involves assessing various risks for their overall risk impact. Compute overall risk by multiplying a risk's probability of occurring by its impact. Risk assessment tables and risk assessment matrixes can be used to compare various overall risks, thus assisting the risk manager is determining which risks require plans for managing them.

Chapter

3

COMPREHENSIVE RISK MANAGEMENT

In 1829, the architect Alexander Jackson Davis (1803–1892) wrote in his diary:

> [Ithiel] Town…is so pleased with my drawings in perspective for the Connecticut Capitol, Hartford City Hall, and Sachem's Wood [all three buildings designed by Town], he kindly proposes an Association to practice Architecture professionally in New York, opening an office at 32 Merchants' Exchange for the transaction of business.(30)

To architectural historians Davis's entry is significant because of two phrases he used: "practice Architecture professionally" and "for the transaction of business." Ithiel Town's (1784–1844) proposal to Davis was groundbreaking for early 19th-century America. The two of them would form a partnership, start a design firm, and practice design as both a profession and a business. Because the idea was new, it was filled with both adventure and risk. The architectural historian Roger Hale Newton and biographer for Town and Davis summed it up this way:

> It has the sound of newness and tinge of adventure... like setting up a balloon service from New York to Mars, or some such fantastical notion...I like to call the firm of Town and Davis approximately the first—or one of the first—of any consequence, duration, and influence to be established in the United States—the first "Association to practice professionally opening an office for the transaction of business."(31)

Although Charles Bulfinch is considered to be the first professional architect in America, he actually supported himself and his family as Police Superintendent of Boston and later as Commissioner of Public Buildings for Washington, D.C. Although Bulfinch worked as an architect, his financial security came from the salary he received as a civil servant, not from his work as an architect. Town and Davis had no such safety net. They set up a design office and hired employees, fully expecting to be paid for the professional design services they rendered. To accomplish this, they had to manage the risks and unknowns that threatened their "first—or one of the first" design practices in the United States.

In the early 19th century, there were no schools of architecture in the United States. The first did not open until 1865 at the Massachusetts Institute of Technology. There were no American architects with formal architectural educations. The first was Vermont-born Richard Morris Hunt (1825–1895), who moved to Paris as a teenager with his affluent parents in 1843, where he attended the École des Beaux Arts, returning to America in 1855. There were no professional societies in the United States either—no organizations to offer suggestions or guidance for setting up and running a professional design practice. The American Institute of Architects (AIA) did not form until 1857, although it was called the New York Society of Architects at the

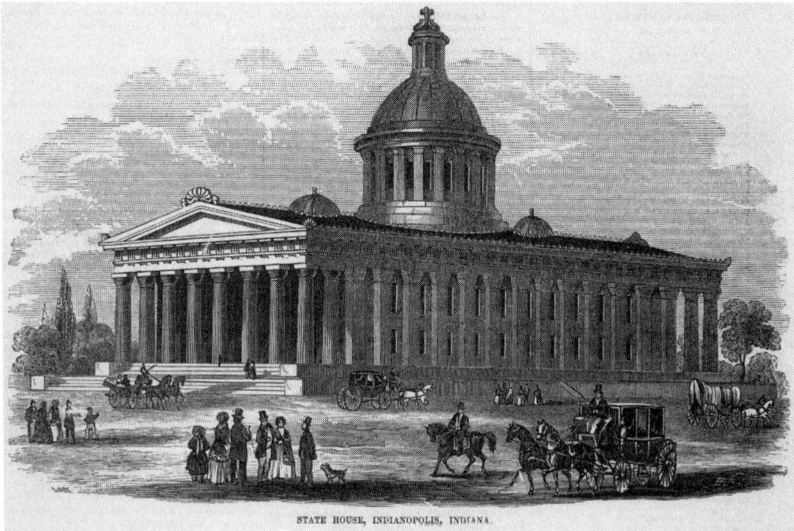

FIGURE 3.1 *The Indiana State House (Indiana State Capitol), Indianapolis, Indiana. Designed and constructed 1831–35. Town and Davis architects.© CORBIS.*

time of its inception. Richard Morris Hunt was among its 13 founders. In early 19th-century America, design was practiced by the self-taught and those who learned from the self-taught. It was within this risk-filled environment that Town and Davis began their design practice.

NARROW VERSUS WIDE VIEW

Today, some design professionals believe that managing risk is only a small portion of the practice and business of design. After all, isn't that why design firms have professional liability insurance and lawyers in the first place? The problem with this narrow view of risk management is that there are many risks associated with the practice of design that cannot be managed by insurance brokers or attorneys. These risks relate to the creative design process, the production of construction documents, the management and training of employees, the marketing of design services, and the host of other activities associated with running a creative and vibrant professional design practice.

The skills required to manage these risks are not found among insurance brokers and attorneys. They are found in the skills, experience, passions, and creativity of design professionals. Design professionals require an all-encompassing, comprehensive view of risk management, one that fully addresses the many risks that weigh on the collective art, science, business, and every day practice of design.

If Town and Davis had held a narrow view of risk management would the "first or one of the first" design firms in the United States have ever gotten off the ground "like setting up a balloon service from New York to Mars"? It is doubtful.

The firm of Town and Davis functioned much like a modern design practice. The principles divided up responsibilities. Town's reputation helped the firm secure commissions for various institutional and monumental projects including the Connecticut State Capitol in New Haven, the Indiana State Capitol in Indianapolis (see Figure 3.1), and the Federal Customs House in New York City. Town was a driving influence in the development and popularity of Greek revival architecture in early 19th-century America. Davis was a superb drafter, artist, and designer or "architectural composer" as biographer Newton called him.[(32)] Davis oversaw the firm's housing projects, designing scores of them over the life of the firm. Together their expertise and efforts complemented each other. Newton's comment about Town and Davis applies to modern design firms as well.

> [It is] difficult to separate members of a firm into distinct professional or artistic entities. I do not attempt to attribute any building to either Town or Davis, singly, during the years of their partnership...I assume that each commission represents their combined talents, in some degree.[(33)]

To compensate for the lack of standards and education, Town moved much of his personal library—over 11,000 books at the time of his death—to the office, making it available to the firm's employees. In addition to managing design projects and the office, Town and Davis actively trained their employees in the art, science, and practice of design.

Like modern design firms, Town and Davis had to make sure that enough work was coming in to support the mouths they had to feed. They juggled project schedules and their personal schedules (Town loved to travel). They responded to the inevitable problems that cropped up during design and construction, and entered into agreements, satisfied the needs and expectations of their clients, and oversaw the construction of their projects. In short, they managed the various risks associated with running a professional design practice. All the while, they produced impressive revivalist architectural designs and buildings that forever left a mark on American architecture.

Like Town and Davis, modern design firms need a broad, comprehensive approach to risk management. Unlike Town and Davis, however, the modern design practice does not have to learn everything about managing risk on its own. While trial and error learning can be rewarding, it can also be painful. It is safer and faster to learn as much as possible from the mistakes and successes of others—apply the successes and avoid the mistakes. To do this consistently requires a risk comprehensive management plan.

COMPREHENSIVE RISK MANAGEMENT

A careful look at the sources of risk listed in the previous chapter suggests a method for managing them. Note that some of the sources of risk on the list are related to running projects, others to employee issues, while still others are associated with the design firm and its organization and management structure. This suggests that multiple methods or plans might be needed to manage all sources of risk. Indeed, that is the case: Many subplans combine to form an overall risk management plan.

Suppose a firm develops a plan for managing its projects and produces a project management manual for its project managers to use. The manual could include proven standards, guidelines, and procedures that the firm knows have been successful on past projects. If used faithfully by project managers, project-related risks could be reduced. Another manual could address common employee concerns and interests and give fair, consistent, and legal answers to the many questions that employees ask time and time again regarding their rights and privileges. An employee manual could explain the firm's benefits package, employee review and promotion procedures, continuing education opportunities, and the firm's bonus and profit–sharing plan. It could provide the information employees like to know in order to feel appreciated and at home within an organization. A well thought-out employee manual could go a long way toward reducing the risks associated with employee dissatisfaction and misunderstandings.

Reasoning along these lines, it turns out that the 15 sources of risk listed in Chapter 2 can be effectively managed by developing half a dozen or so separate plans. Together, the plans form the basis of a comprehensive risk management plan. The plans and the sources of risk they address are:

- Firm business plan: Sources of risk addressed in a business plan are organizational in nature. They are also financial, insurance-related, and contractual. They relate to client type, project type, and external sources of risks. The business plan also establishes procedures and responsibilities for the coordination of the risk management efforts of the other plans.
- Employee relations and management plan: This plan addresses employee-related sources of risk. The plan generally takes the form of an employee manual.
- Office procedures plan: This plan addresses technical performance, quality control, and technological and work

product sources of risk. The plan generally takes the form of an office reporting and procedures manual.

- Marketing plan: This addresses the risks associated with marketing, also called business development. It also addresses certain organizational, contractual, consultant, financial, client, and project type sources of risk.
- Project management plan: The risk addressed in this plan are associated with managing projects, which include project organizational risks as well as technical performance, project financial, contractual, quality control, consultant, bid and award, and construction phase sources of risk.
- Claims response plan: This plan addresses the risks associated with the prevention and response to legal and financial claims against the firm, including those relating to contractual, quality, and insurance sources.
- Quality assurance plan: The risks addressed in this plan pertain to project management, quality of work designed and quality of contract documents produced, technical performance, bid and award, and construction phase sources of risk as well as external sources of risk.

The marketing, project management, claim response, and quality control plans are discussed in more detail in Chapters 4 through 7. The business, employee relations and management, and office procedures plans are discussed in more detail in the following sections.

Note that the plans have considerable overlap in risk management responsibilities. Because there is so much overlap, overall coordination of the plans is essential or else risks may be missed or just moved from one area of responsibility to another, instead of being managed. To avoid this, comprehensive risk management must be part of a firm's overall business plan.

For example, suppose that a firm has a project management plan, but no quality control or comprehensive risk management plans. Suppose the firm's project management plan requires project managers to run their projects on budget—a very reasonable requirement for a project management plan. Suppose a project manager (PM) decides to compromise on quality in order to keep the project from going over budget and does not have the project's final documents checked for quality before sending them out the door. If there is no quality control plan and no procedure for checking project work, a PM may very well do this. In actuality, this happens more than anyone would care to admit. At first, the project manager might appear successful, because he or she successfully managed to keep the project's design costs within budget. However, the true costs of the project may be hidden. It might not be until much later that perhaps a contractor finds the mistakes. It is only then that the project manager, the firm, and the firm's unhappy client learn the project's true costs. Consequently, risk was not averted; it was simply moved from one risk-source category (project management) to another (risks during construction). In addition, the risk may have increased as a result of moving it. Borrowing a phrase from a wise design professional I once worked with, "All that happened was a rearrangement of the deck chairs on the *Titanic*." Without a comprehensive risk management plan, risks can easily be moved around instead of managed, just like deck chairs.

The plans listed above are important risk management tools. If they are properly thought-out and documented, they address all the sources of risks related to the practice and business of design. They lay out procedures for managing risks and assign different individuals within the firm the responsibility of managing different risks. In addition, they engage everyone in the firm in the day-to-day exercise of risk management.

Although the design firm's upper management needs to be involved in the preparation of all the plans, it is helpful to place a single person in charge of developing and managing many of them. In some cases, a committee may develop plans. Even so, the committee needs a chairperson to lead it. Following is a practical and common-sense delegation of plans and risk management.

Delegation of Plans

Business Plan. The business plan needs to be developed by the firm's upper management. Others may have input, but it is ultimately t he responsibility of the firm's owners or major stakeholders—those with the greatest vested interest in its success. One of the firm's principals is often put in charge of overseeing the development of the business plan. There are risk management portions of the plan that may require expert advice from an outside consultant or consultants. There are many consultants that specialize in helping principals of design firms write business plans. While consultants can advise and help, responsibility for the plan still rests with the firm's management.

Employee Relations and Management Plan. This plan is developed by the firm's human resources (HR) officer with assistance from the HR staff. Because the plan must comply with local, state, and national labor laws, it is wise to review the plan with an attorney versed in labor law.

Office Procedures Plan. Some firms assign responsibility for this plan to the firm's chief technical officer, someone responsible for overseeing the firm's technical procedures. Others assign midlevel managers. This could include office managers and discipline or department heads.

Marketing Plan. This plan is developed by the firm's marketing or business development director. Marketing group leaders, market segment leaders, and marketing staff are responsible for following the guidelines and procedures outlined by the plan.

Project Management Plan. The plan should be developed and managed by one of the firm's principals. Project managers are responsible for implementing its procedures on their individual projects. Some firms place a person in charge of all project managers, a super-project-manager sometimes called a manager of projects (MoP). In this case, firm management may delegate responsibility for managing the project management plan to the MoP.

Claims Response Plan. Often this plan is developed and managed by the firm's in-house attorney. If the firm does not have an attorney, then it must be someone intimately familiar with the firm's insurance policy and its provisions. Once the claim response plan is developed, project managers must be made aware of its guidelines, procedures, and reporting requirements, because it is frequently the project managers who are on the front line when a claim or potential claim rears its head.

Quality Assurance Plan. This plan is developed by the firm's chief technical officer or the quality assurance officer. For some firms this may be the same person. The chief technical officer or quality assurance officer is responsible for implementing the plan and auditing it on occasions to ensure that it is being followed. Project managers are responsible for the quality of their projects and, consequently, following the firm's standardized quality assurance/quality control procedures for all project-related work.

THE BUSINESS PLAN

If you set off driving a car with no road map or destination in mind, you run the risk of going around in circles, running out of gas, or driving off a cliff. It is important to know where you are going and why. That is the purpose of the business plan—to define the firm's destination and figure out how to get there. It states what the firm wants to achieve, develops a road map for achieving it, sets a time schedule for achieving it, and establishes mileposts along the way for measuring progress.

The mere act of writing the plan is rewarding. It forces management to think about and research the firm's desired goals and strengths and weaknesses for achieving them. It forces management to consider the firm's organizational structure, the kind of services it wants to provide, and desired project and client types it wishes to serve. The business plan is a feasibility study that assesses the firm and optimizes its chances of accomplishing what it desires.

New firms and old firms both benefit from business plans. Even veteran drivers reach their destinations faster, safer, and with fewer wrong turns when they have a road map.

A business plan starts with a vision—a mental image of what the firm should be like in a reasonable length of time, for example three to five years. The clearer the vision, the better the chances are of achieving it. The vision is like a picture, a character sketch that includes the firm's future physical characteristics like number of employees, dollar volume of business, services provided, number of offices, kinds of projects, types of clients, and so on. It also includes nonphysical characteristics such as design philosophy, firm culture, and reputation. The purpose of visioning is for management to dream about and finally agree on what the firm should be like.

Frequently the exercise begins with management brainstorming and making a list of the firm characteristics that members of management would like to see. The list is then whittled down to qualities that all members can agree to embrace and support.

After the vision is established, a common next step is to summarize the firm's purpose in a sentence or two, commonly called a mission statement. For example, a mission statement might read: "To provide a full range of structural engineering services to a select list of architectural firms that specialize in government and public buildings throughout the western United States." Note that the example has some specificity. It identifies, in a general way, the type of services, client-type, project-type, and geographic area in which the firm wants to practice. Another example of a mission statement: "To provide multidisciplinary design services to relationship-based clients who recognize the value of design of high quality, permitting us to build a high-profile design practice and attract outstanding employees that share our core values of excellence in design, innovation, collaboration, client-focused service, and professional integrity." Note that the latter example forms a run-on sentence, which is very common for mission statements. More importantly, however, it references employees. Recognizing the importance of employees in a mission statement is valuable because employees will, undoubtedly, read the mission statement at some point. Design firms are service-based organizations and, consequently, they are only as good as their staff. Attracting and keeping good employees is paramount to the long-term success of design firms. Firms that understand this and acknowledge the importance of employees in their mission statements send a clear message that they care about employees and their careers. The second example also addresses the firm's values. Values govern a firm's operations and business conduct, and affect a firm's working relationships with its clients, employees, and society at-large.

After the mission statement, the next step is to determine the firm's objectives. Objectives are the results the firm needs in order to achieve its three-to-five-year vision. After listing objectives, strategies must be developed to meet the objectives. In order to determine if the strategies are working, benchmarks must be established to measure progress toward the objectives.

The next step is execution of the business plan, setting it in motion. This is done through a series of programs, plans, and procedures that implement the strategies. For example, a common goal for design firms is to improve profitability, which is essential to the long-term survival of a design firm. To achieve this goal, two common objectives are: improve the utilization rate (the percentage of time charged to projects) and reduce cost overruns on projects. Benchmarks for monitoring the effectiveness of the strategies could be: improve utilization rate by 1 percent per year for the next three years and reduce the overall dollar amount of cost overruns on projects by 1.5 percent per year for the next three years. If the strategies are working, the utilization rate will slowly increase and the cost overruns on projects will slowly decrease. With benchmarks established, it is possible to measure how successful the strategies are and make adjustments if the rate of improvement is not tracking according to schedule. If the strategies are working according to plan, by the end of three years, the firm's utilization rate will have increased 3 percent and the dollar amount of project cost overruns will have decreased by 4.5 percent. Assuming that all other factors that effect profitability remain the same, the firm's profitability will have undoubtedly improved.

It is important to remember that the vision, mission statement, objectives, strategies, benchmarks, and programs are interlinked. They must be logically connected. Implementing a strategy to achieve an objective that supports the firm's mission statement and vision only works if the program, benchmarks, strategy, objective, mission statement, and vision are rationally connected, forming a logical chain. In addition, there must be enough detail in the links or else their inter-relatedness will not be clear and consequently not very strong. If the vision is "just make money," there is not enough specificity in the vision to develop concise objectives for achieving the vision. If just making money is the vision, then why have a design firm in the first place—why not a stock brokerage or investment banking company? It is very unlikely that when today's design professionals decided years ago to become architects, engineers, or planners that their vision was to make gobs of money. "Making money" is not a strong enough vision. There must be more to it.

Like all businesses, design firms must manage risk. If they don't, they probably won't stay in business for the three to five years necessary to achieve the vision. If none of the programs that implement the strategies, achieve the objectives, and satisfy the mission statement and vision, include managing the risks that threaten the firm, then the business plan is missing a vital component. Its absence might very well lead to the firm's demise.

The exercise of writing a business plan often takes many weeks or months. It is a living document. On a regular basis, often yearly, the business-planning process is revisited and the plan is fine-tuned, updated, and revised. New strategies are developed and new benchmarks established to measure progress.

There are various ways of organizing a business plan. Websites offer suggestions and outlines for business plans. They are worth visiting for ideas. (*Planware.org* is an example.) Online business plans, however, tend to be generic in nature. Design professions are unlike other businesses in many ways. The following is a suggested outline for a business plan for design firms:

- Introduction or executive summary
- Vision
- Mission statement
- Management team
- Services offered
- Target clients and/or project types
- Assessment of the competition
- Financial objectives, strategies, and benchmarks (includes insurance and the claims response plans)
- Marketing or business development plan
- Operations plan (includes the project management plan, quality control plan, and office reporting and procedures manual)
- Personnel (includes the employee manual)
- Metrics for monitoring the plan
- Conclusion
- Appendices

EMPLOYEE RELATIONS AND MANAGEMENT PLAN

This plan frequently takes the form of an employee manual. Some firms combine the employee manual and office procedures into one larger document. For the purposes of this discussion, we'll treat them separately.

Sole proprietorships probably don't need an employee relations and management plan—by definition they have no employees. However, as firms grow in size, the importance of communicating firm policies to employees increases. Some employment policies are dictated by local, state, and national laws and must be judiciously followed. Straying from them greatly increases risk. It is wise to consult with an attorney or a professional human resources consultant regarding the various legal requirements that affect employers.

Other employee policies and procedures are designed to produce consistent expectations and patterns of behavior among employees. These include policies regarding office hours, use of the firm's resources, moonlighting, professional development, and a host of other issues that concern both employees and employers. A comprehensive employee relations and management plan reduces the risks associated with treating employees improperly, unfairly, or inconsistently. In addition it addresses many of the benefits or "perks" that keep employees happy and productive, making them more likely to stay with the firm.

Employees are a design firm's primary asset—the firm's bread and butter. Key employees with desirable skills are essential to a firm's well-being and survival. Keeping them requires compensation and benefits such that the employees see their futures intertwined with the firm's.

Design firms spend a lot of money every year training new employees. Many employees stay for a while, become valuable assets, and then suddenly decide to leave. The arduous process of finding a replacement for a valuable employee begins. When key employees leave, it creates holes in the organization that must be filled. It is expensive to find and train new employees.

Design firms with lower turnover rates have an advantage over those with higher turnover rates. Firms with fewer employee turnovers experience less risk to their projects and, therefore, to their longevity as a firm compared to those with higher turnover rates.

To control the risk of employee turnover, one key is to provide alternative career paths. Firms make mistakes in two ways. The first mistake is to reward good employees by promoting them to positions they are not well-suited to perform. Just because someone is a good designer, does not mean, for example, that he or she will be good at managing others. Another mistake firms make is to create organizational structures that only promote those who are willing to jump or change career paths. For example, common career path jumps are to business development, office management, or project management. To keep good employees, all good employees must be able to see career path alternatives that apply to them and are attainable. In recognition of this, some firms have identified career paths for excellence in design and special technical skills. These career paths allow good designers and those with special technical skills to move up in the firm within their areas of interest and expertise.

Design firm managers need to recognize employees that take on additional challenges or do an excellent job. Awards and bonuses are an excellent way to provide this recognition. Not only do employees who receive the bonuses appreciate them, those who don't see that firm management cares about and recognizes good employees, which encourages them to do good work, as well.

To this end, design firms prepare employee manuals that explain their employee recognition programs as well as other policies and benefits that serve and help employees meet their long-term goals. Employee manuals generally include the following:

- Overview of the firm: It may include a brief history of the firm, a statement of the firm's philosophy, and the firm's organization chart.
- Employee benefits package: This section includes an overview of the firm's group insurance benefits, including health care and dental insurance plans, life insurance,

and long-term disability insurance plans. It explains who is covered by the plans, such as employee, spouse, and/or children. It explains the firm's sick leave, holiday, vacation, jury duty, and family leave policies.

- Compensation procedures: This section includes a general discussion of salaries, when they are reviewed, whether standard and formalized employee reviews are conducted, and when they are conducted. It explains promotion procedures, the company's overtime policy and/or incentive compensation policy, and the firm's view of extended leave of absences. It explains whether or not the firm pays for the employees' state licenses and dues for memberships in professional organizations.
- Professional development: A discussion of the firm's career path opportunities and continuing education policies is included in this section. For example, it discusses whether the firm pays for employees to attend seminars and whether the firm pays for the employees' time spent at these seminars. It explains the company's policy regarding research, and the publication of articles, presentation of papers to professional organizations, and so on. It also discusses the firm's policy regarding continuing education tuition reimbursement.

OFFICE REPORTING AND PROCEDURES PLAN

This plan generally takes the form of an office procedures manual. It explains the company's procedures, rules, and employee reporting requirements. The manual is important because it standardizes procedures, letting employees know how they are supposed to behave in the office and while representing the firm outside of the office. It tells employees what the firm's various expectations from its employees are, including the firm's principals. The office reporting and procedures plan explains:

- Personnel procedures: These include expected employee behavior and conduct, as well as the use of firm equipment, sexual harassment policy, dress code, drug abuse policy, outside employment, use of the firm's vehicles, maternity and paternity leave, mail, email, telephone, and so on.
- Administration and resource support: This includes the names of the firm's human resources specialist(s), computer support personnel, and legal support.
- Accounting procedures: This explains the firm's invoice approval process, use of company credit cards, expense account forms and documentation, purchase of books and reference materials, documentation of phone calls, and standardized reporting forms for meetings.

If the office procedures manual is not well written, it can sound oppressive. I remember working for a small design firm that lightened up the procedures manual with cartoons drawn by one of the firm principals.

CHAPTER SUMMARY

Design professionals should adopt a comprehensive view of risk management. This is achieved by developing and then managing seven plans. They are:

1. Business plan
2. Employee relations and management plan
3. Office procedures plan
4. Marketing or business development plan
5. Project management plan
6. Claims response plan
7. Quality assurance plan

Chapter

4

RISK AND BUSINESS DEVELOPMENT

Architectural historian Henry-Russell Hitchcock singled out the 19th-century architect Henry Hobson Richardson (1838–1886) as a seminal figure in modern architecture. In his 1936 book *The Architecture of H. H. Richardson and His Times,* Hitchcock likened Richardson to a proto-modern architect—at once "the last great traditional architect" and the first "reformer" who inspired and "helped [Louis] Sullivan."(34) According to Hitchcock, if one traced the path of modern architecture backward, it eventually arrived at the door to Richardson's design studio.

Adding to Hitchcock's thesis, the architectural historian James Marston Fitch wrote:

> If the churches, store buildings, and railroad stations which Richardson was then designing [during the 1880s] impressed the public with their outward appearance of order, organization, and plain good sense, it was because Richardson so largely understood the complex organisms of modern capitalism—the huge department store, the large and worldly metropolitan church, the modern

> library—he was called upon to [design]... He was...capable of grasping the large outlines of the problem before him, of analyzing the forces involved, organizing them into a coherent and workable plan.[35]

It is the plan that separates modern architecture from traditional architecture. Traditional architecture began with the elevations, whereas modern architecture starts with the plan. Hitchcock and Fitch identified Richardson as modern because Richardson began with building function and from this developed "a coherent and workable plan."[36]

Legends grow around seminal characters from history, and Richardson is no exception. True or not, there is a story about Richardson that is apropos to a discussion about business development. Taking an interest in her son's design practice, Richardson's mother once asked him, "What is most important in the practice of architecture?" Richardson answered, "Getting the first job." Expecting a somewhat different answer, his mother then asked, "Well yes, but what is most important after that?" to which Richardson answered, "Getting the next job."[37] Another version of the story has Richardson explaining that, "there is not one thing that is most important, there are three, getting the next job, getting the next job, and getting the next job."[38] Regardless of which version one prefers, the story stresses the importance of business development in the practice of design.

There are a number of things that design firms can do to increase the chances of getting the next job—and there are things they should not do. While selling is good, there are risks associated with overselling, such as making promises that cannot be kept.

FIGURE 4.1 *The three steps of the process of getting the next job are like three parts of a funnel. Marketing is associated with the wide end. Business development is a focusing activity and is associated with the narrowing body of the funnel, and sales is the narrow end that deposits new projects into the firm's backlog of projects.*

MARKETING, BUSINESS DEVELOPMENT, AND SALES

Before embarking on a discussion about risk and business development, it is worth clarifying a few terms—"marketing," "business development," and "sales." Some firm principals say they have "marketing staff." Others say they have "business development staff" and still others say they have "sales staff." Some call their staff all three, using the words interchangeably. The terms do have slightly different meanings.

Marketing is general in nature. It is an ongoing activity that takes place when no specific job or client is identified. Before a duck hunter can shoot a duck, the hunter must first attract the ducks. Marketing is that process. Marketing consists of the general things that design firms do to enhance their image and promote themselves, such as sending out brochures or newsletters. It involves writing magazine articles, having booths at trade shows, and/or presenting papers at conferences. Marketing activities describe the firm and what services it provides to clients and potential clients.

Business development (BD) is a more focused activity. It is meeting one-on-one with clients to discuss their particular needs and upcoming projects. Business developers (BDers) track specific clients and specific projects.

Sales is the process of landing a specific project or a specific client. The project is in sight; it is the firm's to win or lose. Sales is winning the job, negotiating the details, closing the deal, and booking the win.

As the descriptions suggest, marketing, business development, and sales are different steps of the same process—the process of getting the next job. I once worked with a business developer who likened the process of getting the next job to a funnel. At the top of the funnel (the wide end) was marketing, where the funnel began to narrow, he called it business development, and at the narrow, focused end it was sales.[39] (See Figure 4.1.)

Because it is both wordy and clumsy to say "the process of getting the next job," when referring to the overall process, the terms "marketing" or "business development" are used interchangeably in the following discussion as stand-in names for "the process of getting the next job." The funnel analogy—I accredited it to a "business developer." His title was actually "regional director of business development" as I recall, but it could just as easily have been "regional director of marketing," "regional director of sales," or some similar title. Like the overall process and the names of the parts of the process, titles are also used interchangeably.

APPROACHES THAT DON'T WORK IN THE LONG RUN

Before discussing business development processes that work consistently and thus decrease the risk of not getting work, it is worth looking at three approaches that usually don't work in the long run. These approaches work once in a while under special circumstances, but they cannot be relied on as the foundation of a sustainable business development effort. The three approaches are:

1. Chase anything that moves approach
2. Silver bullet approach
3. Silver spoon approach

Chase Anything That Moves

My wife and I have two cats. We have had one of them considerably longer than the other. When we got the second cat, he was just a kitten and he chased anything that moved. He would chase his tail and even his own shadow. When he did, the older and wiser cat looked on with distain, as if wondering, "What kind of nut is this guy?"

Chasing everything that moves takes a lot of energy, which is no problem for a kitten, but not so for design firms. Energy for a design firm takes the form of money and time. Design firms don't have an endless amount of either. A design firm can't afford to chase every job it hears about. Doing so wastes precious time and money. It also zaps BD staff enthusiasm because it reduces the probability of success.

A mail order or telephone solicitation business can run profitably on a 3 percent success rate. These businesses only need a small percentage of wins to keep going. A friend of mine once worked as a telephone solicitor. He told me that he learned not to get depressed when he was hung up on all the time. He rationalized that for every time it happened, he was just one phone call closer to that 1 in 33 that would say "yes." Just imagine how depressing and inefficient it would be for a design firm professional to submit 33 proposals or have employees go to 33 interviews to win just one job!

A design firm with only a 3 percent success rate will not stay in business long. To succeed frequently, design firms need to be disciplined and selective in deciding which projects to chase. They should only chase those jobs that they have a good chance of winning and are qualified to do, instead of chasing everything that moves.

The Silver Bullet Approach

The Lone Ranger relied on silver bullets to do his talking. He kept his six-shooters full and carried a backup arsenal of them on his holster belt. He fired silver bullets left and right and when the smoke cleared, he won every gun fight—and he did it all without ever losing his hat!

Design firms aren't like the Lone Ranger. They don't have an infinite supply of silver bullets, meaning they cannot rely on special favors such as a few right words from the right friend in the right ear and "presto" the project is theirs. Although this does happen, these situations are rare and they cannot form the basis for a sustainable business development plan. No firm has that many silver bullets.

The Silver Spoon

Some smaller design firms are built on the privileged connections of their founding principals. As long as the principal is able to tread on his or her fortunate and elite connections, the firm can succeed. For these principals and their design firms, the world is presented to them on a silver platter, giving them the freedom to nibble as they choose, as a child fed with a silver spoon. Low and behold, for these privileged souls, things just fall in place. Most design firms, however, cannot rely on the birthright of their founder to survive. They must develop sound business development plans in order to succeed.

The architect Phillip Johnson (1906–2005) was born into a wealthy family. Never for want of money, he tried his hand at history, philosophy, art history, architectural history, museum directorship, and journalism before finally concentrating on architecture. In 1939, at the age of 33, he returned to Harvard to study architecture. He designed and began building his famous glass house in New Canaan, Connecticut, for himself, on his own property, as his master's thesis. He finished it 10 years later in 1949.

In the early 1950s, Phyllis Lambert, another silver-spooner and daughter of the president of Joseph E. Seagram & Sons, took a few art history classes in college. Her father gave her the assignment to find an architect for his Seagram Building in New York City. Lambert contacted Johnson who introduced her to his friend Mies van der Rohe. Van der Rohe was not licensed in the state of New York but Johnson was, so Johnson became involved in the design of the Seagram Building, assisting van der Rohe with the detailing. Of course Johnson had talent, and it was this asset and not his money on which his fame was ultimately built, but his silver spoon certainly fed him opportunities.

Building a marketing strategy around a silver spoon cannot work for the vast majority of design firms. Design professionals cannot pick their parents. Nevertheless, they can choose to develop a sound and sustainable business development plan.

BUSINESS DEVELOPMENT BASICS

The three previous approaches rely on luck and special circumstances, which, by definition, nobody can control. To survive in the long run design firms must take control of their fate. They must make a business development plan and follow it. The plan must be procedural, repeatable, and sustainable. Others must be able to learn it. Principals and staff members move on and may leave the firm. When this happens, new employees must be able to learn the plan, follow its procedures, and continue winning work. The plan must have metrics for monitoring its success. Without some way to measure progress, it is difficult to tell whether the plan is working.

Developing a procedural, repeatable, sustainable, and learnable business development plan begins with a frank assessment of the firm itself. What is the character of the firm? What are its ambitions, its goals? What is the business plan? Without a candid self-assessment it is difficult to develop and organize a successful effort for getting work. Honest answers to the following questions are the first step in developing a business development plan:

- How is the firm organized?
- What are the firm's strengths and weaknesses?
- Should we chase projects or clients?
- If projects, what size projects?
- If clients, which clients? Commodity or value-added? Novices or experienced? New or repeat? Public or private? Is the client financially stable? Does the client pay on time?

How Is the Firm Organized?

Some design firms are organized in the studio format, others in matrix formats. As I discussed in *Project Management for*

Design Professionals, as a general rule, matrix organizations tend to be strong project manager organizations with well-defined responsibilities for project managers. Studios tend to be more autocratic and project management responsibilities are often shared between a principal and a project architect or engineer.[40] The organization of the firm affects the size, number, variety, and complexity of projects that a firm can successfully work on simultaneously. This in turn affects the business development effort.

Design firms organize their BD efforts differently. Some firms are set up with full-time BDers, others are not. Firms with full time BDers are often called seller-and-doer firms because they have separate sales staff (sellers) and operations staff (doers). Other firms are organized as "seller/doer" firms, meaning that the seller is also the doer. Seller/doer firms do not have full-time BDers. Instead principals and project managers do marketing when they are not working on projects.

While there is no right or wrong organizational approach, there are differences between the two approaches, which are discussed in more detail next. Whichever approach is used, the business development plan must understand, respect, and work with the firm's particular organizational model.

What Are the Firm Strengths and Weaknesses?

I once worked for a small design firm that was set up in the studio format. The principal liked to call his practice a general architectural practice. I thought this was curious because we always seemed to be working on the same kinds of projects, not that the projects we worked on were the only ones we chased. We went after all sorts, but we only won the kind with which we had experience. In reality, we didn't have a general architectural practice; the principal only thought we did. Design firms must know themselves, their strengths, and their weaknesses. They must recognize what kinds of clients they work well with and the types of projects they like and are successful performing—and what to avoid. Business development plans should build on the firm's strengths and not waste effort chasing projects or clients that play into the firm's weaknesses. Continually chasing projects that the firm has no experience doing or doesn't do successfully only siphons off BD resources that could be better spent chasing projects that the firm knows how to win.

Chase Projects or Clients?

Some firms go after projects. They focus their BD effort on finding particular types of projects and then position themselves to be selected. The client is of secondary importance because it is the project type that drives the BD effort. Other firms go about getting work the opposite way. They forge relationships with clients. They court those with whom they have had successful working relationships. They slowly and carefully expand their small circle of preferred clients. These firms focus their business development efforts on convincing good clients to think of them for all their projects, regardless of type. Other firms do a bit of both. They only look for certain types of projects with certain kinds of clients. All three approaches can be successful, providing the BD effort develops procedures that value and support the approach.

What Size Projects to Pursue?

Size matters: small multidisciplinary projects are often just as difficult to manage as large multidisciplinary ones. The difference, of course, is the size of the fee. Smaller projects have smaller fees. Depending upon the nature of the design firm, some project sizes are too large for the firm, some too small, and others just right. Design firms need to know what size is best suited for them. One size does not fit all. In many respects, it takes almost as much effort to win a small job as it does to win a large job. Consequently, many firms establish a threshold, a project size or fee amount below which they do not chase. I recall the president of a large design firm with thousands of employees telling a group of us that he used to celebrate a million dollar win and that it kept him happy for weeks. "But now," he said, "I don't even ring the victory bell for a million dollar win."

Which Clients to Pursue

Commodity or Value-Added? Some clients buy design services as if they are buying pencils or office equipment. These are called commodity-type clients. For commodity-type clients, price is a big factor in their selection process. Sometimes it is the only factor. Commodity-type clients generally have standard types of projects, nothing terribly special or unique about them. Some design firms excel at these types of projects because they have learned how to do them inexpensively, meaning they have learned how to spend fewer hours at lower hourly rates than their competitors have.

Other clients have projects that are one-of-a-kind, perhaps high-profile projects like a museum or city hall, and are looking for value-added design firms. Value-added clients are interested in finding the best design firm for their project. They select design firms based on qualifications. Price is a secondary consideration and often is not discussed until after the design firm is selected.

Working for these two client-types can be completely different experiences. For the design firm, controlling design costs is critical to the success of commodity-type projects. Budgets are usually tight, so design firms cannot afford to investigate very many alternatives. The right solution has to come pretty quickly. Projects move fast and schedules are often as tight as the design fee.

Value-added clients are generally not as concerned with hourly rates as commodity-type clients are. Value-added clients are willing to pay for a thorough investigation of alternatives in pursuit of the best design for the project. Firms that are used to working for one type, often find the transition difficult working with the other type. A firm's overhead structure, quality assurance program, and project management procedures often favor one client type over the other. Successful business development plans are different for the two client types, because one client favors price over qualifications and the other qualifications over price.

Novices or Experienced? Some clients have regular facilities management groups staffed with their own design professionals. These clients build projects frequently and have considerable experience working with design professionals and constructors.

Yet novice clients build infrequently, in some cases, only once. For example, a husband and wife deciding to build their one-and-only dream house are novice clients. They have little or no experience working with design professionals or contractors. Governmental bodies can also be novices. The Scottish Parliament Building, completed in 2004, and the Sydney Opera House, completed in 1973, are two good examples. The governmental bodies that oversaw the design and construction of these buildings had little experience managing large, unusual, and complicated design and construction projects. The Scottish Parliament Building went 1,100 percent over budget and the Sydney Opera House 1,400 percent. In the soul-searching afterward, members of the Scottish Parliament admitted that they knew very little about designing buildings.

Forging successful working relationships with novice clients requires time, patience, and the willingness to teach. Some design firms are up to the task, but some are not. Design firms working for novice clients must be willing to spend the time necessary to educate their clients regarding the design and construction process and advise them of the risks involved. Design firms must be willing to make decisions for them, decisions that experienced clients would normally make themselves and might even resent design professionals making for them.

Experienced clients know the kinds of design professionals they want. It is much easier and less risky to work with clients who know and appreciate the design process and can keep problems in proportion. Things frequently go wrong, but generally they can be worked through if the client does not overreact and think the sky is falling. Experienced clients generally have the expertise to evaluate the completeness and technical merits of submittals. They frequently have standards and procedures they expect design professionals to follow. They understand the process, costs, and risks associated with design and construction.

Design firms need to figure out which kind of client, or what mix of client types, is most appropriate for their particular practice and expertise.

New or Repeat? Inherent in working for novice clients is that they need design services rarely, perhaps only once. The problem with making novice clients the mainstay of a design practice is that the marketing for design services must start over fresh from project to project. As discussed next, there is a long course and a short course for getting new work. Pursuing new clients requires playing the long course over and over, which is more expensive and takes longer to play. Getting work with past clients shortens the course considerably because it is less expensive to play and doesn't take as long. Because this is the case, design firms that concentrate on doing work only for novice clients put themselves at a disadvantage compared to the majority of design firms. According to a 2007 survey by the American Institute of Architects, 58 percent of architectural firms' total billings come from repeat clients and only 42 percent from new clients. The percentage for repeat clients jumps to 70 percent for firms with 50 employees or more,[41] suggesting that if a design firm plans to grow, it better figure out how to increase its repeat client workload.

Work from repeat clients takes much of the marketing effort out of the hands of the BDers and puts it in the hands of operations and project managers. Doing a good job is the best business development there is. Consequently, the best business developers in the design firm are not BDers at all; they are exceptional project managers and designers. Some might argue that this is how it should be; because design is a profession, the quality of services provided should be the primary business development tool.

Public or Private? Private clients run the gauntlet from individual home owners, to small businesses, large businesses, private and publicly held corporations, and multinational corporations. Private owners can choose their design professionals however they want. Depending upon their sophistication and preferences, the selection process can take many forms. Private owners can select their design professionals based on qualifications, by proposal, by cost, or by bid. They can even just hand a design professional work if they want to.

On the other hand, public owners such as municipalities, local and state governments, and federal agencies must follow standardized rules for procuring design services. In general, these rules require the public owner to hire based on design skill and qualifications, rather than on cost or fee alone. Another generalization: public owners are more likely than private owners are to deal with design professionals fairly when procuring design services. The public owners' selection process generally involves more steps than that of private owners and is consequently more expensive. The process usually includes submission of the firm's qualifications. The public owner compares the qualifications of numerous design firms and prepares a shortlist of firms to continue on in the process. The short-listed firms—usually three to five firms—prepare for and attend interviews where they present their qualifications and discuss their design approach for a particular project. The public entity selects the most qualified based on an objective, or somewhat objective, selection process. The most qualified firm submits a fee proposal for the work and negotiations ensue. If the public owner and the design firm cannot come to an agreement, the public entity moves on to the next most qualified design firm.

Is the Client Financially Stable?

Design firms want to get paid for their services, so they need to do their homework and investigate new clients. Developers can pose special problems. Generally, firms work with developers they know and with whom they have had successful track records. Developers often get the idea that if they enter into an agreement with a design firm and the project does not pan out, it is acceptable not to pay the design firm. The rationalization goes something like, "Well, we didn't make money, so the design firm doesn't either." Some design firms work for developers in an arrangement where they share in the success of the project, which means that if the project tanks, the design firm does not get paid. Recall the story of Charles Bulfinch from the introduction. A design firm still has staff to pay, so it needs to consider the risks involved before getting into relationships with developers that the firm might regret later. Some firms have policies that they simply do not work for developers.

Does the Client Pay on Time?

Some clients drag out payments as long as possible. Generally speaking, most design firms like to get paid within 30 days after invoicing. Frequently this doesn't happen. According to PSMJ, the average accounts payable time frame is around 70 days.[42] Many design firms have agreements with their subconsultants that state that the subconsultant does not get paid until the design firm gets paid. For design firms that take the role of subconsultant, this frequently means a considerable delay between when the work is performed and when the firm gets paid. Some design firms pay consultants faster than others. Some are willing to pay smaller invoices from small consultants before they get paid. It is worth trying to find out the time lag in getting paid. If the design firm's role is that of subconsultant, the client must pay the prime design firm before the design firm pays the subconsultant. Monies have to change hands a few times before the subconsultant gets paid. For this reason, some design firms shy away from being second-tier consultants or subconsultants.

THE BD PLAN

Business development plans need to be written down. They are too important to be left floating around in the heads of a few key firm principals. If the plan is not written down, it is difficult or impossible to explain it to others in the firm. As a result, what the plan is and what is expected from various firm members will remain vague and out of focus. At a minimum the plan should address the following:

- Business development mission
- Organizational structure
- Marketing goals
- Marketing approach and procedures

Business Development Mission

The business development mission should be based on the firm's vision. Recall that the mission is a general summary of purposes and desires. Because this is the purpose of the business development group, and not the firm as a whole, the mis-

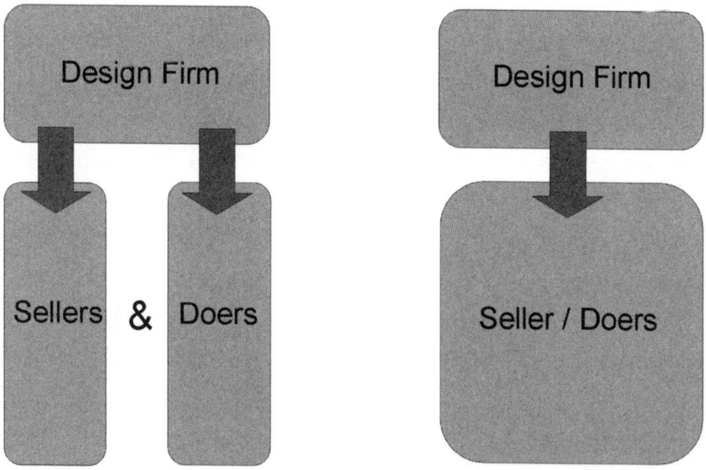

FIGURE 4.2 *The Seller-and-doer and the seller/doer business development models.*

sion is slightly different from the firm's overall mission. BD's mission is to support the firm's vision by bringing work through the door that is consistent with that vision.

Organizational Structure

It is important to know how the business development effort is organized. Its organization goes a long way toward understanding how it is supposed to work. As previously discussed, there are two fundamentally different ways to organize the BD effort—the sellers-and-doers and the seller/doer models (see Figure 4.2).

The sellers-and-doers model separates the business development effort from operations. This results in a business development organization that is hierarchical. In general, full-fledged BD departments include the following. There is someone in charge, such as a principal or a director of marketing. There are BDers, those responsible for getting work. Frequently, BDers are divided by market sector. Depending upon the firm, the sectors may be by client type or by project type or a mixture of the two. There is production staff, sometimes called marketing support staff. These are the people that write, prepare the artwork, and assemble the qualifications or proposals and get them out the door on time. This group frequently includes technical writers and graphic artists. It is also essential that somebody in the production group be an excellent editor, speller, and grammarian.

Seller-and-doer firms tend to be a bit less efficient at business development than seller/doer firms are. According to PSMJ's *A/E Financial Performance Survey* for 2006, seller-and-doer firms spend 5.4 percent of net revenues on business development compared to 3.2 percent for seller/doer firms. Nearly all the difference is attributable to labor costs.

The tradeoff is that with the seller-and-doer model someone is always thinking about business development, whereas with the seller/doer model that may not be the case when the firm is very busy with project work. In addition, the seller-and-doer model opens up more career path opportunities within a design firm. Providing various carrier alternatives for staff is a plus that may very well offset the higher business development costs of the seller-and-doer model.

Town and Davis is an example of a seller/doer firm. Both partners were heavily involved in getting work for their firm as well as designing projects. H. H. Richardson's office was also a seller/doer firm with staff that contributed to the design. Richardson set his office up in an atelier, or workshop, format where he was simultaneously the leader, designer, teacher, and heavily involved in the design of all office projects. The firm of Burnham and Root was more of a seller-and-doer firm. Burnham was the seller and John Root was the designer or doer. When Root suddenly died in 1891, the firm was heavily involved in planning the 1893 Chicago Columbian Exhibition. Without Root, Burnham was a seller without a doer. He quickly found another doer, the architect Charles Atwood. Atwood is credited with the design of dozens of

BUSINESS DEVELOPMENT PROJECTIONS

Project	Estimated Fee	Probability	Adjusted Fee
XYZ Project	$250,000	80%	$200,000
BCD Project	$200,000	90%	$180,000
UVW Project	$150,000	100%	$150,000
EFG Project	$300,000	75%	$225,000
RST Project	$800,000	30%	$240,000
		TOTAL PROJECTIONS	$995,000

FIGURE 4.3 *Procedure for tracking business development projections. Estimated fees for prospective projects are multiplied by their estimated probabilities of winning to equal the adjusted fee.*

buildings for the Columbian Exhibition. After the fair, Burnham and his firm became heavily involved in the American Beautiful Movement and prepared a number of city planning master plans for cities such as Washington, D.C., San Francisco, and Chicago. Burnham relied heavily on the architect and planner Edward Bennett for much of the planning.

Marketing Goals

Unlike the firm's vision, which includes three- to five-year goals, marketing goals are generally annual. Typical marketing goals might be:

- To sell *X* amount of dollars of services this year
- To achieve that goal, market sector A needs to sell *X* amount; sector B, *X* amount, and so on

- To reduce our business development costs as a percentage of gross revenue by *X* percent this year

There are many other possible goals. Note, however, that the listed goals here are specific and can be measured. It is possible to monitor progress toward achieving them. BDers and the firm management can analyze data and determine if results are progressing on schedule. Without the ability to measure goals, it is impossible to know if the business development strategies and procedures are working, need to be adjusted, or if some more radical fix is required.

Some marketing goals are not as easy to quantify. For example, a design firm may decide it wants to improve the quality of its proposal writing or the graphic presentation of its proposals. The firm may want to expand into other project-type markets. Although steps toward achieving these goals are more difficult to quantify than the goals listed, some method of measuring progress toward them should be identified. Some possibilities are:

- To hire a technical writer to assist with proposal writing
- To hire a graphic artist this year to help us improve the graphic quality of our proposals
- To send our BD production staff to business development seminars that specialize in 1) writing winning proposals and 2) improving graphic presentation of proposals
- To identify two additional market areas for our services and develop a strategy for moving into these market areas

Marketing Approach and Procedures

These explain the process that the business development group will follow in order to meet the marketing goals. The approach or approaches and procedures must be tailored toward the specific project and/or client types and goals. If every BDer is free to choose his or her way of pursuing work, it will invariably lead to inefficiencies, duplication of efforts, and lost opportunities because important things will fall through the cracks.

Some approaches to business development are considerably less risky than others are. For example, pursuing projects where the firm has a proven and positive track record is considerably less risky than going after projects that the firm has never done before. In addition, pursuing work from past clients where the relationship was good involves less risk than doing work for new and unknown clients. Of course, these two strategies can lead to stilted design practices that become too risk free and, therefore, uninteresting or unchallenging. Most design professionals like projects that present some sort of challenge. Challenging projects stretch the design firm's capabilities and expand its portfolio of work, making it easier to successfully pursue more types of projects and work with different kinds of clients. Many design firms make it a practice to seek out projects that expand their capabilities, but are related closely to their current experience and skills so that risks can be more easily managed.

It is important to track projects and clients. Procedures for tracking include writing memos to document visits and phone calls and keeping a tickler list that reminds BDers when to again contact potential clients with a phone call or visit.

It is also good to know the approximate dollar volume of potential work. If a design firm's marketing goal is $1 million in fees for the year, it would be nice to know the potential dollar volume of the work being tracked. A procedure for doing this is to list all the projects that are being tracked, the expected fees for each project, the likelihood of each project going ahead, and the chances of the design firm being selected for each project. Figure 4.3 on page 97 illustrates the process.

Of the projects listed, note that the largest has the lowest probability of winning. Also note that it dramatically affects the dollar amount of the projected sales. In fact, it skews them into a picture that is likely way too rosy.

I recall a BDer who listed numerous projects he was tracking. Most of them were large projects with small chances of success. Nevertheless, if you list enough projects with small chances of winning, you can come up with some pretty hefty sales numbers. When he first joined the firm, upper management was pleased with his sales projections, but as months of no sales turned into yearly quarters with no sales and eventually into two years with no sales, upper management changed their opinion dramatically. He lasted with the design firm just a little over two years.

To avoid painting unrealistically rosy pictures, many firms do not count projects with low probabilities that BDers are tracking. Some firms set a threshold of 50 percent probability when calculating revenue projections for the coming year. Others set the limit at an even higher probability, say 75 percent.

THE LONG AND SHORT COURSE OF BUSINESS DEVELOPMENT

I used to travel a lot. I managed two design offices separated by 400 miles. I lived near one of the offices and had an apartment near the other. When I was away from home, I'd sometimes play golf after work. Depending upon the time of year, I could get in anywhere from 6 to 14 holes before dark. Rarely could I squeeze in 18.

I played normal-sized par-72 courses that play 6,000-plus yards. During the winter (winters in Southern California are not severe) I played a short 18-hole golf course, a so-called executive course made up of par-threes and short par-fours. Because it was short and had lights, I could play at nighttime and often got in 18 holes.

I much prefer playing longer golf courses because the round takes more time, the course is more challenging, and I get to use all my clubs. When it comes to business development, however, I like short courses.

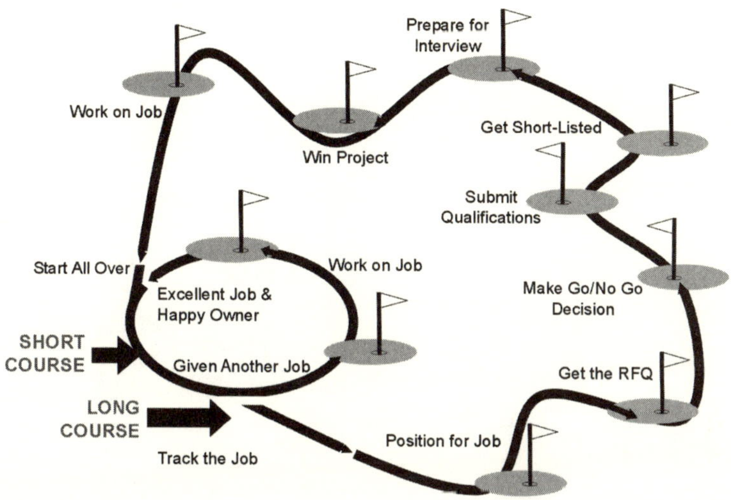

FIGURE 4.4 *The long and short course of business development.*

There is a long and a short course in business development. The long course includes all the general steps of business development. It starts with promoting and positioning the firm to win work. It includes developing a list of clients and projects worth tracking and pursuing, and cultivating those relationships. It includes receiving requests for qualifications (RFQs) or requests for proposals (RFPs) and deciding whether to respond to them. It requires putting together proposed design teams for projects, preparing the qualifications and proposal submission data, getting short-listed, preparing for and attending interviews, making the interview presentations, winning the projects, ironing out the details of the agreements, and then doing the projects. The hope is that when a particular project is finished it is done well, finished on time and budget, and the client is pleased. The process then starts over with the next project.

This is the long course of business development, and it is played over and over hundreds of times. It requires a firm to hone all of its skills and use every club in its business development bag.

FIGURE 4.5 *Trinity Church, Boston, Massachusetts. H. H. Richardson architect. © Bettmann/CORBIS.*

As every BDer knows, it can get very pricey playing long and challenging business development courses all the time.

Alternatively, there is a short course to business development. It is less expensive to play and not nearly as time consuming because it leaves out a number of BD steps.

The short course works on the premise that if you do a good job on a project, you can remove many steps in business development in getting the next one. The short course starts with a successfully completed project. In the process of working with the client on a successful project, a good working relationship is forged. The client, pleased with both performance and the working relationship, gives the design firm the next job sole source. A fee is negotiated and the firm is off and running on to the next project. There is no waiting for the RFQ or RFP, deciding whether to go after the particular job, preparing the submittal, getting short-listed, being interviewed, and hoping for a win. Instead, all the firm has to do is its job well on the last project.

See Figure 4.4 on page 101 for a depiction of the two courses of business development.

A handful of H. H. Richardson's buildings serve to illustrate the importance of the short course. In 1873, Richardson began work on Trinity Church in Boston (see Figure 4.5 on page 102). In 1874, he moved to Brookline, a suburb of Boston, to oversee its construction. There he developed a relationship with Frederick Lothrop "F. L." Ames, the only son of Oliver Ames II, a rich and prominent merchant and manufacturer of shovels. The architectural historian William H. Pierson, Jr. wrote:

> It could not have been long after the architect's arrival in Brookline that the two came together and began the intimate and productive friendship that would culminate in one of the most remarkable architectural achievements of nineteenth-century America.[(43)]

Pierson is referring to the assemblage of buildings in Brookline all designed by Richardson for his client, Ames and family. The relationship helped shorten the business development course for Richardson. Richardson did not win his first commission with Ames based solely on friendship. The Trinity Church was a great success.

The construction of Trinity Church was completed in February 1877. One month later, Oliver Ames died and left $50,000 in his will for the construction of a new free-lending library. In September 1877, F. L. Ames hired Richardson. Richardson did an extraordinary job. The library is an amazing example of the picturesque movement in architecture. Ames was so pleased with Richardson's design that the BD course shortened even more. In early 1879, the Ames family gave another commission to Richardson: an auditorium, the Oakes Ames Memorial Hall. Longwater Gate Lodge followed this project in March 1880, and then in November 1881, the Ames family extended another commission to Richardson, the Old

Colony Railroad Station, followed by the Gardeners Cottage at North Easton in March 1884, located near the aforementioned Gate Lodge. One well-done project, Trinity Church, led to many others. The lesson for design firms: Doing a good job shortens the business development course.

I recall a conversation I once had with the principal of an MEP firm. His firm did excellent work and he was always busy. The architectural firm I worked for used this MEP firm over and over. I asked him how he did business development. He answered with a smile, "I don't." Of course he did, it was just that he spent almost no money doing it. He did his business development working on projects and providing excellent service to his clients. He did have a brochure of sorts that was two lists, followed by a few letters. One list was sorted by client type and the other by project type. Following the lists were three or four letters of recommendation from satisfied clients. After the last letter there was a sentence that read, "More available upon request."

GO/NO GO FORM

Before investing a lot of time and money pursuing a particular project, it is a good idea to decide whether it is worth the effort. To this end, many firms employ a formalized decision-making process called Go/No Go. The process includes a standard list of questions, and frequently the process uses weight factors to prioritize the importance of the answers. Depending upon the relative weighted importance of the answers, firms decide whether to put their effort toward winning the project. Following are some of the major factors considered.

Have We Been Tracking the Project?

When the project was off on the horizon, did the firm know about it and keep track of it, or did the firm just happen to learn about it in an announcement sent to many firms? Pursuing projects that have been tracked is generally more successful than just chasing after anything that comes along. This takes discipline and a restrained approach to marketing. Of course there are always projects that just happen by accident. While it's nice to get these, and many firms have stories about them, the serendipitous approach is hardly a sound business development strategy. Generally successful design firms track upcoming projects well in advance.

Is the Project Pre-Sold?

This can be good news or bad. If the design firm has tracked the project and positioned itself so the client decides to give it to the firm, then the selection process is little more than a formality. However, pre-sold projects can mean the opposite. It could be pre-sold to another design firm and then it is probably not worth pursuing. I say *probably* because sometimes it still makes good business sense to go after it. Some clients expect its preferred list of design professionals to "play the game," so to speak. Many clients who have active facilities management groups work with a number of favorite design firms. It is nice to be a favorite, but it is also an obligation. These clients expect their favorites to submit proposals on their projects. The client picks the firm it wants. The trick, of course, is to know which turn is yours. On these proposals, put in a good effort and spend the time and money necessary to make a thorough proposal. When it's not your turn, go through the motions and don't spend too much money.

Many years ago, I attended an interview where the project was supposedly pre-sold. I didn't know it at the time, and so I spent a lot of time preparing for the interview. Our team practiced and we made a very good presentation. The firm that was the favorite dogged it—they thought they had already won the project, so they didn't do a good job in the presentation. The firm was very surprised when we won instead. The lesson here is that pre-sold isn't necessarily a sure thing.

Is It an Existing Client?

Additional work with an existing client has definite advantages. The design firm knows what the client is like. If the design firm liked working with the client, chances are it will like working with the client again. If it didn't like working for them, then maybe the firm learned enough about how to work with the client to make the next project a better experience. I have heard more than one principal of design firms comment, "We fire bad clients." Bad clients can be those that do not pay invoices in a timely manner or change their minds frequently yet are unwilling to augment the design firm's fee for making the changes. Bad clients can be confrontational, always on the look out for someone to blame or take advantage of, or are litigious, frequently threatening to take legal action if things don't go the way they expected. If a client doesn't work out, then move on. For good design firms there are more fish in the sea. Before taking on a new client the design firm should do a little homework. The firm needs to decide if the client falls within a core market area or core client type. Is it a good client to work for? It is worth talking with other design professionals about their experiences with the client. Find out what you can about potential clients before deciding to work for them.

Do We Understand Client Needs and Expectations?

The design firm should do its homework before responding to a request for qualifications or proposal. Read the qualifications or proposal request carefully to ascertain what the client wants and whether this meshes with the services the design firm has to offer. Determine if the project is in a core strength area or if it's a stretch. Some clients like to work with firms nearby; others don't care. Make sure you know what this client thinks. Some municipalities like to work with local businesses. Maybe you are nearby, but not in the same city. Find out if that makes a difference. Determine if this will be a one-time client (many clients build only once). Decide if this the kind of client the design firm wants.

Are We Qualified?

Is the firm qualified and are the proposed design team members qualified? Sometimes firms claim experience, but those individuals with relevant experience are no longer in the firm! Other times, individuals have experience but the firm does not. A well-qualified firm requires both. Many clients have told me that when they interview design firms they are really looking for a project manager. Determine whether or not the design firm's best project manager is available for the job.

Do We Have Staff Available?

Is the project manager available for the project or is he/she already overbooked? Some firms are notorious for overbooking good project managers. Once they get the project, they must scramble in order to perform it. If staff is stretched too thin, design teams can miss deadlines, perform poorly, and make mistakes. This might not bode well for future work with the client.

What Is the Selection Process?

Is the selection process fair or is it tilted in such a way that it favors some other design firm? Everyone wants a level playing field, unless it is canted in his or her favor. Is the selection process time consuming and therefore expensive? How many are being short-listed? Some clients short-list and interview too many firms. Three or four is common, five is getting high, and six or more is too many. If six are short-listed, five will walk away disappointed. There is no point in one of them being your firm.

Do We Know the Selection Panel?

Knowing who is on the selection panel is a big help because it gives the firm insight into the panel's individual likes, dislikes, preferences, and concerns. One step better is to have good relationships with members on the selection panel.

What Are the Likely Marketing Costs and Fee?

Firms should figure out in advance how much it will cost to win a job and then determine if it is worth it. For example, suppose it takes $10,000 to win a $100,000 job. Chances are the profit on the project isn't much more than 10 percent—that's $10,000. If you win, the best case is you'll get back what you spent chasing it. Under some circumstances that is acceptable. Maybe it is a market area you are trying to break into, or maybe it's a client you are trying to hold onto. Maybe times are tight and you need the work to keep key staff busy. All of these are good reasons to pursue a project that is break-even at best, but know in advance the likely costs so an informed decision can be made.

Do We Have Marketing Staff Available?

Does the firm have a proposal writer and a proposal manager available? Is the graphic artist overwhelmed with work? If there is nobody available to manage and write the proposal, then it may not be possible or in the firm's best interest to pursue the project.

GO/NO-GO FORM PROJECT NAME XYZ PROJECT

DECISION CRITERIA	YES/NO/ * NOT SURE	WEIGHT FACTOR **	SCORE
HAVE WE BEEN TRACKING THE PROJECT?	0	5	0
IS THE PROJECT PRE-SOLD TO US?	1	5	5
IS IT AN EXISTING CLIENT?	0	5	0
DO WE UNDERSTAND CLIENT NEEDS?	1	3	4
ARE WE QUALIFIED?	2	3	6
DO WE HAVE STAFF AVAILABLE TO DO THE WORK?	2	3	6
DO WE HAVE BD STAFF AVAILABLE?	2	3	6
DO WE UNDERSTAND/LIKE THE SELECTION PROCESS?	0	4	4
DO WE KNOW SELECTION PANEL MFMBERS?	0	4	4
DOES POTENTIAL PROFIT EXCEED MARKETING COSTS?	1	5	5
		TOTAL	40

* YES = 2, NOT SURE = 1, NO = 0

** WEIGHT FACTORS 1 TO 5. 1 = NOT IMPORTANT, 5 = VERY IMPORTANT

FIGURE 4.6 *Go/No Go form.*

Organizing Your Questions on a Go/NoGo Form

The previous questions can be standardized on a Go/No Go form. Figure 4.6 is an example of this form. Note that for this particular project the total score was 40 out of a total possible score of 100. The total score will depend upon the project, number of decision-making criteria, and weight factors. Through experience, the firm determines a baseline score—scores above the baseline suggest that the project is worth pursuing. Scores below it mean the firm will not pursue the project. In the Figure 4.6 example, it is probably not worth chasing the XYZ Project. The design firm had not been tracking it, the firm does not know the selection process, nor does it know anyone on the selection panel, and it is not a project with an existing client. These are all good reasons for passing on a particular project.

PROJECT MANAGER'S ROLE IN BUSINESS DEVELOPMENT

Regardless of a firm's organization, when it comes to pricing a fee proposal, the project manager's input is very important. Without it, the project's financial success may be at risk.

Over the course of my 30-plus years of experience, I have witnessed many projects whose fee was determined by BDers. Once the fee was determined and the contract signed, the project was given to a project manager to perform. Among PMs, projects like this are called over-the-transom projects, meaning that they mysteriously appear in the PMs office without any notice. Some BDers rationalize that design services have a market value and because they do, there is no point in figuring out how much it will cost to perform the work: the fee is the fee. Some clients encourage this attitude. Federal agencies, for example, have guidelines regarding fees as percentages of the cost of construction. Currently, the final design fee for many federal agencies, called Title II services, cannot exceed 6 percent of the estimated construction cost. Direct costs, certain consultants, programming, conceptual design, and other professional design services are exempt from the 6 percent, but straight design work and the preparation of construction documents must be done for 6 percent of the estimated construction cost. Those in business development can do the arithmetic easily, so why bother to ask the PM for input?

I recall being short-listed for a project for a university. With the congratulations letter telling us our interview time and date, there was a paragraph that explained the university's fee-structure policy for the project—6 percent of the cost of construction with an additional 2 percent because the project was a remodeling project. The letter told us what the maximum design fee would be. If we were still interested, we could show up for the interview. We decided—rationalized—we could do the work for the fee. We needed the work! We showed up for the interview and made our presentation. During the questions and answers session, we were asked if we could live with the maximum fee amount. We said we could. We won the project and proceeded to lose our shirts.

The project manager's role is to help the business development staff prepare a realistic fee for the project, a fee that will actually allow the design firm to do the required work. The project manager who will be responsible for the work has a vested interest in providing a fee that is as realistic as possible. The PM is in the best position to interpret the scope of work and appreciate all the steps involved and the level of difficulty of the project. The PM knows what disciplines need to be involved and what information is needed from the different disciplines in order to put a realistic fee proposal together. In multidisciplinary firms, the disciplines may be in-house. For single discipline firms, it is necessary to get fee input from consultants. In many cases, it may involve a combination of the above. Once the BDer receives fee input from the PM, the number may need some massaging. If so, it should be done in consultation with the project manager. When the number is finally turned in, it should have buy-in from the PM. In general, PMs do not appreciate BDers who hand them projects with fixed fees that are losers right from the start. Projects like this are no fun to manage.

RISK AND PROPOSAL WRITING

As the saying goes, "The devil is in the details." In proposal writing, it is in the wording. Certain words are riskier than others are, and because they flow off the pen so easily, their potential impact is often overlooked.

The point of proposal writing is to sell the firm without overselling it. Risk is increased when the proposal claims that the firm can provide services it can't perform or that its liability insurance won't cover. Many times a firm never meant to make the claims that it made. The words just poured out so easily that before anyone is the wiser, the words are in the proposal. The firm is selected and now the client expects the design firm to live up to its promises—promises that it cannot keep.

If a firm oversells itself, it can come back to haunt the firm later. There are a few phrases that crop up frequently in proposals that carry with them potential risks.

"Our firm will prepare complete and detailed drawings and specifications." When the firm falls short, and change orders arise during construction because something on the drawings or in the specifications was wrong or omitted, the client may become upset that the firm did not live up to its words. In reality, no firm can produce a perfect set of construction documents. Therefore, firms should not say that they can.

Words like *"insure," "ensure,"* and *"assure"* crop up in proposal writing all the time. These words can pack a wallop. They claim something that the design professional cannot do. Rather than say, "We will assure that…" it is better to say, "We will make a reasonable effort to…"

"We guarantee a successful project." Because many elements that contribute to a project's success lie beyond the control of the design professional, there is no way that the design professional can guarantee success. In addition, design professionals provide services, not products. They are not designing a mass-produced gadget that has had the benefit of months or years of testing before thousands of copies are made. Each design project is unique—a prototype, so-to-speak. It will have bugs in it. That is the nature of making one-of-a-kind designs. Because design professionals are providing services, they are held to a different standard of care than manufacturers of products are. As discussed in Chapter 7, guarantees are big mistakes because they blur this very important distinction. Design professionals should not guarantee or warrant anything. It is better to say something like, "We are committed to this project and will make every effort to meet the scope, schedule, and budget requirements."

"We have highly trained professionals." This phrase or variations of it are very common. "Highly trained" suggests that the client

can expect the firm to perform better than the normal standard of care because the phrase says that the firm has better-than-normal design professionals. It claims that the design firm's design professionals are somehow trained beyond the normal professional level. It is very unlikely that this is true. It is better to say, "We have experienced design professionals," or "Our design professionals have an average of *X* years of experience."

"Always" is a word to avoid. So are *"must"* and *"should." "Best"* is a good word to avoid and so is *"obvious,"* as are their opposites, *"never," "none,"* and *"zero."* These words carry with them the expectation of absoluteness.

"We will attend all necessary or required meetings." This sentence leaves the firm open to attending an unlimited number of meetings. It is better to establish a cap for meetings, even if it is somewhat arbitrary. Nobody knows how many meetings are necessary or will be required to design a large project. It is better to say something like, "We will attend up to four public meetings," or "We agree to attend a total of four public meetings."

"We will attain a variance for the project." The variance process is often paved with subjectivity. It has twists and turns that the design professional cannot control. Sometimes one irate citizen can thwart the hard work of many design professionals. It is much safer to say, "We will assist the client in attaining a variance, with the client's understanding that it is not possible to know with certainty the outcome of the variance process."

"We will conduct a field survey and gather all existing information." A design firm cannot know in advance that it will gather all information. Chances are that something will be overlooked. Things hidden in walls, attics, and crawlspaces can't be discovered until demolition occurs. Buried abandoned pipes and concrete structures are found all the time. Nobody knows one is there until a backhoe hits it. Is the firm planning to do demolition work and pothole the entire site in order to collect "all existing information"? It is better to say, "We will collect readily available information," or some similar phrase.

CHAPTER SUMMARY

Don't chase every project that comes along and don't rely on the silver bullet or the silver spoon approaches to business development. There is a long course and a short course to business development. The long course has the following steps:

- Promoting and positioning the firm
- Developing relationships with potential clients
- Tracking upcoming projects
- Getting an RFQ or RFP
- Analyzing the RFQ or RFP and making a Go/No Go decision
- Submitting a statement of qualifications or proposal
- Getting short-listed
- Preparing for an interview
- Making a presentation
- Winning the job
- Negotiating a fee
- Doing the project
- Starting the whole process over again

The short course shortens the process considerably and has the following steps:

- Doing a good job on a project
- Owner is happy with the outcome
- Owner selects the firm for another project
- The process starts over again

A sound business development plan is sustainable, repeatable, learnable, and has metrics for monitoring. It requires:

- Knowing the firm's strengths, weaknesses, and viable market areas

- Knowing the firm's BD structure—is the firm a seller-and-doer or seller/doer?
- Developing a strategy for deciding what projects to chase—a Go/No Go process
- Deciding whether to chase clients or projects
- Identifying projects in advance
- Being disciplined and only chasing those projects that are part of the plan
- Understanding your competition—what they're doing that works and doesn't work
- Assigning different individuals responsibilities for different parts of the plan

A business development plan includes:

- Mission statement
- Marketing goals
- Marketing strategy
- Marketing group's organizational structure
- Responsibilities of marketing staff
- Marketing procedures
- Market areas or sectors

Chapter

5

RISK AND PROJECT MANAGEMENT

Many years ago, the vice president of a large multi-office, multidisciplinary design firm told me that the project managers in his office did not work for him. "It's the other way around," he said, "I work for them." He explained that the success of the office hinged on the success of the projects, and their success depended upon the project managers. Whether projects ran smoothly, were of high quality, pleased clients, and made money depended upon the skills of the project managers. "My job is to support the project managers, encourage them, and smooth out the bumps that get in their away," he told me. He thought of himself as more than an office manager—he was also facilitator, coach, and cheerleader.

Since then, I have had the opportunity to manage a couple of multidisciplinary design offices myself and I've learned firsthand what he meant. Successful projects are the backbone of successful design practices and projects don't turn out successfully by accident. Good project managers make it happen.

PROJECT MANAGER SKILLS

Design firms are service organizations and, as such, they are labor intensive. Most revenue comes from billing project-related labor hours, not billing the costs of materials or equipment used to provide design services. Because project managers manage the projects and design teams, they are crucial to the financial well-being of a design firm. If the project managers do not have the right set of skills, project risk increases because the probability of projects turning out well is greatly reduced. To prevent or mitigate this risk, it is important to understand what skills are needed to manage projects.

Leading creative and well-educated teams of design professionals requires special qualities. A design team is not like a mule team or dogs pulling a sled. Individuals with different levels of expertise and educational backgrounds cannot be lashed together and expected to pull in the same direction. Project managers need leadership capabilities and a balanced set of organizational, design, and technical skills in order to create and sustain the team environment that encourages creativity and cooperation. Because project managers frequently manage more than one project at a time, they must be able to change their mindset quickly and redirect their attention from project to project, which are likely to be at different stages in the design process.

Project managers need a thorough understanding of the design process and have a general grasp of the design business. They need to know enough about the disciplines of planning, architecture, and engineering to communicate intelligently and effectively about design and construction related matters with other design team members, consultants, clients, and builders. It requires enough knowledge about design and construction to make the necessary, correct, and timely decisions needed to keep projects moving along productively. Project managers that do not have the right characteristics increase project risk. The characteristics are:

- Desire to be a project manager
- Organizational skills
- Design and technical competency in the project type
- Generalist's attitude
- Ability to make decisions with incomplete information
- Professional and ethical behavior
- Strong communication skills
- Ability to motivate design teams
- Risk manager's attitude

Desire to Be a Project Manager

The best project managers are those who like the role. Happy project managers have high motivation to lead. People with high motivation to lead have favorable attitudes toward authority and strive for positions where they can exercise authority. They enjoy the respect, power, and accolades that come from successfully managing projects. They also enjoy leading, managing, coaching, and teaching others. Combining a high motivation to lead with strong organizational, design, technical, and communication capabilities yields a well-qualified project manager that enjoys his/her job. Because people tend to do best at what they like, project risk decreases when those who like leading and managing others lead projects.

Organizational Skills

Project managers have to plan their work and the work of others. They have to break down projects into logical and manageable pieces or tasks. Because design projects are frequently multidisciplinary, the work of one discipline often affects and is dependent upon the work of others. Interrelated tasks must be thoughtfully scheduled and budgeted. Metrics for tracking progress must be established. On a regular basis, project managers need to monitor progress. If things are not going according to plan, adjustments must be made. All of this requires exceptional organizational skills. Project risk decreases when the project manager possesses strong organizational skills.

Design and Technical Competency in the Project Type

I have been told that a good businessperson can manage a bottling company, a shoe factory, or a department store. The skill set required to manage different types of businesses is supposedly transferable. Whether or not this is true, the design and technical skills required to manage one type of design project are not necessarily transferable to other types. Managing a design project requires an understanding of the specific key design and technical factors that bear on the success of that particular type of project. For example, managing the design of a bridge involves many design and technical decisions that are completely outside the area of expertise of an interior designer. The opposite is also true. A bridge designer would be an unsuitable choice to manage a 200,000-square-foot tenant improvement project with a tight budget and schedule, because the decisions that bear on tenant improvement projects are likely to be completely foreign to bridge engineers. Project managers must have a basic understanding of the design and technical issues that must be resolved for the project to be a success. They must know the order in which the parameters that affect the design need to be resolved and have an idea of how and how long it takes to resolve them. They must be able to look at work in progress and judge its completeness: is it 50 percent complete or 75 percent? None of this can be done effectively if the project manager does not have the proper design and technical skills. As a general rule, project managers should be from the lead discipline. If it is a structural engineering project, a structural engineer should lead the project. If it is an architectural project, an architect; a mechanical engineering project, a mechanical engineer. Project risk increases when the necessary skills are lacking.

General List's Attitude

The previous discussion might leave the wrong impression. While project management requires design and technical competency in the type of project, it does not benefit when project managers get caught up in minutiae. Project managers must be generalists. That is to say, they cannot allow themselves to get lost in details at the expense of the bigger picture. Detailed and technically minded people frequently do just that. They focus only on the part of the project in which they have greater interest and expertise. Rather than a narrowly focused view, a project manager needs to rise above the details, see the forest through the trees.

I recall a job interview for an extensive remodeling and addition project for a food processing facility. Our proposed project manager was a structural engineer. Prior to the interview, he visited the existing facility for the purpose of familiarizing himself with the facility and determining what planning, design, and technical considerations would affect the project. A number of factors affect the design of food processing facilities, such as sanitation, food processing work flow, cold-storage warehousing, and on-site truck circulation, to name a few. During the job interview, rather than talking about these factors, he focused on one particular aspect of the project that interested him most of all—an existing concrete beam, and whether it was large enough to take the added load of the addition. The concrete beam was of consequence, of course. Analyzing it and properly sizing it would be necessary during design. However, he elevated its significance beyond its relative importance and downplayed other major factors that affected the design of the facility. The client, no doubt, would have preferred a discussion about these major factors. Incidentally, we did not get that project.

Ability to Make Decisions with Incomplete Information

Design projects are not like the story problems on math tests you took in elementary school. When you read the math story problems, all the information was there. To get the right answer, all you had to do was figure out which numbers to plug into the formulas that you memorized the night before. In the real world of design, decisions must be made without all the information. Frequently there are no formulas. This stymies some project managers. They want to wait until all the information is available and then make the decision. The problem is, in many cases, all the information never comes. The project manager's indecision becomes the decision, and no decision is usually not the best decision.

Recall from Chapter 1 that pragmatic decisions are based on abductive logic, which is educated guessing based on expertise and experience. Project managers who are not willing to make decisions with incomplete information put their projects at risk. When decisions are not timely, budgets can be exceeded and deadlines missed.

I recall a project that included an emergency generator housed in a concrete block building. The generator building was only a small part of a much larger project, which had a very tight design schedule. While the bulk of the project was proceeding nicely, there was indecision on the part of the client regarding how much processing equipment needed emergency backup power. Consequently, the size of the generator was in question, and this affected the size and design of the concrete block building that housed the generator. The project manager postponed working on the generator building as long as she could in hopes that the client would finally decide how much processing equipment needed emergency power. Eventually, the project manager could no longer wait without jeopardizing the project schedule, which she knew was paramount to the project's success. She made an abductive decision that managed risk in two ways. First, her decision kept the project on its all-important schedule, and, second, she kept the client's options open regarding the emergency generator. She instructed her design team to size and design the building for the larger generator. By designing the building to accommodate the larger generator, the client could take a bit longer to make a decision. The worst that would happen would be that the generator building would be a bit bigger than it needed to be, should the owner finally decide on the smaller generator. Prior to the design team embarking on the design of the generator building, she phoned the owner to discuss her decision. The owner was pleased with the decision and, as a show of appreciation, continued to postpone making a decision about the generator until the bitter end.

Professional and Ethical Behavior

Nothing can put a project at risk faster than a project manager who behaves unprofessionally, fraudulently, or dishonestly. The Golden Rule and everything it entails applies to project management: Treat clients and coworkers as you would like to be treated. Not only is ethical behavior important for project managers, it is essential to the longevity of the design professions. In recognition of this, professional organizations such as the American Institute of Architects and the National Society of Professional Engineers have codes of ethics that they expect their members to follow. Above all else, project managers—indeed all design professionals—have an obligation to the general safety of the public. This is why they are licensed. Project managers are often the flag bearer for their firms. As such they must be faithful, honest, and trustworthy representatives.

Strong Communication Skills

Project managers with poor communication skills put their projects at risk. If they are unable to explain issues to clients and design team members, it is unlikely that the issues will be properly resolved. Project managers must be able to communicate orally and in writing, and know how to draw well enough to convey ideas. Design professionals are visual people; they often absorb the meaning of drawings better than words. Many clients, on the other hand, do not read drawings well and rely more heavily on words. Project managers must be able to communicate effectively with both.

One common communications tool is the program statement, which explains the scope of work. In the case of a building, it includes descriptions of the spaces, area requirements, and other pertinent information that pertains to the planning and design of the building. These documents are useful tools for conveying information to clients that do not quickly read drawings. Project managers must be adept at developing scopes of work and communicating task information to team members. They must be able to talk to clients, team members, and consultants about complex technical issues.

Ability to Motivate Design Teams

Design is a team effort. Project managers must have the ability to get the design team pulling together in the same direction—or run the risk that their projects will not be completed on schedule or budget. Without a unified effort projects are less likely to be properly coordinated and this can lead to mistakes. Motivated design teams have enthusiasm for their work and respect for their teammates and the project manager. They work together effectively and coordinate their work. They share the same goals: completing the project on time and on budget while meeting all scope requirements. Motivated design teams are empowered design teams. Empowered design teams have a clear plan and they know what they are supposed to do, when they are supposed to do it, and how much they can spend to do it.

Risk Manager's Attitude

As planner of the project and leader of the design team, the project manager is in the best position to see ahead. Because the project manager communicates regularly with the client, consultants, and project team members, he or she is most likely to recognize potential project threats lurking on the horizon. No one else is in a better position to avoid risks. Consequently, the project manager must embrace the concept of risk management, plan to avoid risks, monitor the results of the actions taken, and make course adjustments as necessary.

WHY PROJECTS FAIL

Projects fail in a variety of ways but for six basic reasons:

1. The project was impossible in the first place due to technological and/or knowledge limitations.
2. One or more project objectives were unrealistic and therefore impossible to achieve.
3. Reasons beyond everyone's control caused project failure, such as acts of nature.
4. It was impossible to succeed because of the way the project was sold.
5. Execution of the project was poorly planned.
6. Project work was poorly executed.

FIGURE 5.1 *Collapsed Tacoma Narrows Bridge, © Bettmann/CORBIS.*

Knowledge and Technological Limitations Increase Risk

Leonardo Da Vinci wanted to fly. He studied birds, drew freeze-frame sketches of them in flight, and made many drawings of ideas for flying machines. He even constructed a flying machine and attempted to fly. His ideas were centuries ahead of his time. Human flight was impossible during the Renaissance because the age lacked sufficient knowledge about fluid mechanics and the airfoil. It also lacked the technology to fly. One of Da Vinci's drawings was for a glider with maneuverable tips. Although the bat-like machine looked like it might be able to fly, it would have been much too heavy to get off the ground if constructed of wood and stretched canvas—the only materials available to Da Vinci.

For risk managers, the lesson in Da Vinci's flying machines is that prototypical projects that are far ahead of the current state of theory and/or technology carry with them unusually high levels of risk. Project risk management is based on historical project data used to help predict future outcomes. If there is no historical data on which to predict future outcomes, then risk increases.

Risk also increases if a project goes well beyond historical precedents. The Tacoma Narrows Bridge in the state of Washington is an example. When it opened on July 1, 1940, its 2,800-foot span made it the third longest suspension bridge in the United States. Its elegant 420-foot towers and narrow two-lane wide road bed were meticulously designed to reduce the bridge's weight and cost. The bridge's designer Leon Moisseiff had impeccable credentials. He had served as a consulting engineer for the Bronx-Whitestone Bridge in New York City, the San Francisco-Oakland Bay Bridge, and the famous Golden Gate Bridge. Almost immediately after its completion, the Tacoma Narrows Bridge experienced problems. Its roadbed moved up and down and swayed to and fro in the steady and often heavy winds typical in the Tacoma Straits. While other suspension bridges experienced sway and vertical movement in heavy gusts of wind—60 mph or more—they all settled down quickly, once the gusts subsided. This was not so with the Tacoma Narrows Bridge. Like a father pushing his child in a swing, the pushing winds of the strait amplified the bridge's swaying motion. Quickly the bridge earned the nickname Galloping Gertie. On November 7, 1940, four months after it opened, Galloping Gertie literally tore itself to pieces in a sustained wind of less than 40 mph (see Figure 5.1). When asked why his bridge failed, Moisseiff could only answer, "I'm completely at a loss to explain the collapse."[(44)]

What Moisseiff did not know, nor did anyone else at the time, was that harmonics play an ever-increasing role in bridge design as the span of the bridge increases and the weight of the roadbed decreases. In addition, important in bridge design is the ratio of roadbed width to span. As the ratio increases the ability of the roadbed to resist twisting forces decreases, much like a slender column has less ability to resist buckling than a stouter and heavier column. The point for risk managers is that when a project pushes beyond the bounds of historical precedence and

FIGURE 5.2 *Scottish Parliament Building Blast Wall. Photo by the author.*

moves into uncharted territory, risk increases. When it does, historical project data may be of limited help in predicting future outcomes, thus increasing risk.

Unrealistic Project Objectives

Projects don't have to fall down to fail. They can also fail if one or more key project objectives are not met. Two common key project objectives are project cost and schedule. The Scottish Parliament Building illustrates what can happen if project schedule and cost objectives are not realistic. When the Scottish Parliament Building opened in October 2004, it was hailed by many architectural critics as a postmodern masterpiece. Nevertheless, it was also mired in the throes of controversy. It opened years behind schedule and cost the Scottish taxpayers 11 times more than its 1999 estimate. During design and construction, its cost and schedule problems were daily fodder for the Scottish press. The building was originally programmed with a total area of 170,000 square feet and an estimated construction cost of $75 million. However, in 2000, its program of spaces was revisited and increased, resulting in a dramatic increase in building area, nearly doubling to 325,000 square feet. The size of the change and its timing—after much design work had already been competed—made it unrealistic to meet two key project objectives—completing on budget and on time.

It is impossible to construct a building of nearly double the area for the same cost. Consequently, the building became considerably more expensive to construct. In addition, the building's postmodern nonorthogonal geometry and its complicated detailing exacerbated the problem. Many details were so complicated they had to be figured out in the field, adding considerably to the cost. When finally completed, the Scottish parliament building was nearly four years behind schedule and cost a staggering $830 million.

Reasons Beyond Everyone's Control

Two additional events added to the difficulty of completing the Scottish Parliament Building on time and budget. In July 2000, the building's architect Enric Miralles suddenly died and three months later Donald Dewar, the First Minister of Scotland and major promoter of the Scottish Parliament Building, also died. Their deaths left the project in turmoil during a crucial period. Without the project's two central characters, it was not possible to make hard and fast decisions regarding the quickly escalating costs of the Scottish Parliament Building. Another event that no one could have anticipated at the project's outset added cost to the project: the 9/11 terrorist attacks in the United States. After September 11, 2001, many additional terrorist-deterrent features were designed into the Parliament Building, including a costly sculptural concrete fortification blast wall along the Royal Mile, designed to resist bomb blasts (see Figure 5.2).

Impossible to Succeed Because of the Way Project Was Sold

A primary task of business development is to convince the client that the design firm is better qualified for the job than any other firm. Having been selected for the project, the next task is to close the deal. This involves negotiating a compatible fee, schedule, and scope of work for the project. If this is not done properly, the project may be a loser from the start.

Many years ago I worked on a large project that involved the design of many different facilities for a large corporation. Because the project included programming and the study of alternative ways of consolidating the owner's facilities it was difficult to define a specific scope of work. Consequently, a time-and-materials (T&M) fee agreement was settled upon, meaning that for every hour the design firm spent, it was paid an hourly rate. (There is more about different fee arrangements in Chapter 7.) In this T&M arrangement, the hourly rate was determined by multiplying an employee's hourly salary rate by an agreed to multiplier that included profit.

It is nearly impossible to lose money on a T&M agreement—nearly. If the multiplier is too low, it will not cover the design firm's overhead costs. The owner's representative was a shrewd negotiator. In need of work for his office, the design firm's office manager agreed to a salary multiplier that was too low, fearing that the owner would take the project (and money) elsewhere. We started working on the project and after a few weeks the design firm's home office figured out that we were losing a few dollars for every hour we spent working on the project. A vice president from the home office confronted the office manager and asked what went wrong. The office manager explained how the negotiations went, saying that the owner's negotiator forced him to agree to the multiplier. The vice president asked, "Forced you? You mean he held a gun to your head?"

Poorly Planned Projects

Projects fail because there is either an ill-conceived plan or no plan at all. In addition projects can be made worse if recovery plans are ill-conceived. The Dome of the Santa Maria in Florence, Italy, had no plan for its completion when first designed by Arnolfo di Cambio in 1296. For generations, it remained unfinished because no one could figure out how to design a dome to cover its enormously wide crossing. It took 125 years to complete the church and the amazing talents of a Renaissance genius, Filippo Brunelleschi (1377–1446), to pull it off.

The 1906 San Francisco earthquake started many separate fires. With little water to fight them, an ill-conceived recovery plan evolved. Recall that recovery plans deal with the effects or outcomes when something goes wrong. Even recovery plans must be thought through or else their implementation may make matters worse instead of better. The idea was to use explosives to level buildings in the paths of the fires. The result was that glowing embers from the explosions started additional fires. At one point as many as 50 fires burned in San Francisco.

Project risk is greatly reduced if the execution of projects is well thought out in advance. Moreover, should something go wrong, it is equally important that recovery plans be well thought through or matters can be made worse.

Poor Execution

The final reason why projects fail is actually the most common. Poor execution is the easiest to control through proper risk management. A number of project execution risk management techniques follow.

RISK MANAGEMENT TECHNIQUES

Managing Risk Before the Project Starts

I remember preparing for a project interview where I was slated to be our firm's project manager. We were extremely busy at the time and working long hours. I wondered how we were going to do the project if we won it. I asked the manager of business development and he told me, "We'll figure out how to do it after we win it."

The business development manager's words notwithstanding, project risk management should begin before the project starts. If the firm has a comprehensive risk management plan, which should include a business development plan, then many project-related risks will be addressed and resolved by the marketing group before a project becomes a project. Projects that are considered very risky or inappropriate for the firm will have already been weeded out. The only projects the business development group should pursue are those within the firm's areas of expertise and that are consistent with the firm's goals, as stated in the business plan.

Various ways of controlling project risk before the project starts include:

- Make sure the firm has the right project manager with the right level of skills and experience for the project. Good project managers are hard to find and are often busy. Find one and get a commitment from him or her for the project. Project managers that are spread too thin put projects at risk.
- Have the project manager and design team participate in determining the fee for the project (as discussed in Chapter 4). Project fees are more realistic when those who will perform the work participate in establishing them. Called bottom up fee budgeting, the process promotes

team "buy-in," which is important in reducing the risk of a project going over budget.

- Examine the project for unusual scope, budget, and schedule requirements. Anything unique regarding any of these three affects the complexity of the project and consequently increases risk.
- Confirm that there is a project designer available. Often the PM and the project designer are not the same person, particularly on larger projects. Like good project managers, design firms frequently overbook their best designers. Designers that are spread too thin put projects at risk.
- Analyze the project's proposed construction contracting method. Determine if it will be design/bid/build, negotiated by the owner directly with a contractor, design/build, or constructed under some sort of construction management agreement between the owner and a construction management (CM) firm. These different methods carry with them different types and levels of risk.
- Determine if the project is politically sensitive or if the community has particular concerns. If so, project risk increases. It can be very costly to navigate the twists and turns and jump the hurdles thrown in a project's path by politics and special interest groups. On occasions design firms hire public relations firms to advise how best to handle sensitive political and community related issues.
- Make sure the owner has realistic expectations. Some owners have unrealistic construction budgets, others unrealistic schedule requirements. The project is destined to fail from the outset unless the owner's expectations can be changed. Be selective choosing clients as well as projects.
- Confirm staff is available. A multidisciplinary project requires a qualified person from all disciplines available to work on it. In addition, make sure that key staff members agree to the design fee and project schedule. If not, the project is at risk.

- Make sure the firm can accept the owner's agreement requirements. Some owners have their own agreements, which are frequently weighted in their favor. Can the firm live with the owner's insurance requirements and indemnity clause? Some agreement clauses are deal-breakers because they carry with them high risk.
- Sell comprehensive services whenever possible. Risk is reduced when design firms are involved with a project from programming through preliminary design, final design, and construction. Sell construction administration services whenever possible—it offers additional opportunity to catch mistakes made during design before they are constructed.

Managing Risk when Setting Up the Project

After the project has been won, it is time to execute it. If work is slow or the project is particularly exciting, it might be tempting to jump in and start designing and burning hours right away. That's risky. Work that is unplanned or uncoordinated is unlikely to satisfy all of the owner's scope and deliverable requirements and runs the risk of not being completed on time or budget.

The first step is to plan the project and break it down into manageable steps so that staff can work on it effectively. To this end, there are a number of project risk management considerations.

Be sure the scope of work is clearly defined. If it's not, it is the project manager's responsibility to define all project objectives so that every team member knows what is expected. Projects with clearly defined scopes of work pose less risk than those where the requirements are not known or are vaguely defined. "Design me a house," is an extreme example of the latter case. It is so vague that it is almost impossible to know exactly what is expected. You don't know what kind of house, how many bedrooms, how many bathrooms, where will it be located, when the owner wants it built, or how much money he or she wants to spend. Beginning design work before knowing exactly what "design me a house" means can be very risky. The chance of hitting on a solution that meets the client's many unstated requirements is very low.

Consequently, one of the essential first steps in project management is to make sure the scope of work is clear, concise, and understood by everyone—the project manager, the design team, and the client. The responsibility to clarify the scope of work falls on the shoulders of the project manager. Project managers who are not up to the challenge increase risk because design teams who embark on projects with vague or incomplete scopes spin their wheels and waste money.

Determine whether or not the entire scope of work be accomplished in-house. Some design firms are multidisciplinary and able to design many projects in-house. Most design firms, however, do not have all necessary professional disciplines in-house. During the project-planning phase, all necessary consultants need to be identified and budgets established for their work.

Assuming that consultants are needed, find out if the design firm has worked with the consultants before. If so, inquire about their work—was it performed well and completed on time? Ascertain whether the consultants meet the owner's or the firm's insurance requirements, if they understand the scope of work, and if they can commit to the project budget and schedule?

Explore if there is a project contingency. When assigning hours to various tasks, it is a good idea to tuck a little money away as a contingency. It is not likely that the project will run perfectly; something always goes wrong. The number and severity of things that can go wrong can be reduced through proper planning and risk management. Nevertheless, it is wise to establish a project contingency. Ten percent is frequently chosen, but this can vary with the project particulars.

Consider whether some of the owner's requirements are subject to change. Some clients aren't exactly sure what they want, and a careful reading of their scope of work reveals this. Other times, it is a conversation with the owner (or owner's project manager) that first suggests to the PM that changes are likely. Frequent changes can increase project risk. PMs must be wary of owners who frequently change their minds. When changes occur, PMs must tell owners of the design costs, construction costs, and project schedule implications of changes.

Research whether the project requires the design professional to invent something or use new or unproven technology. If so, this increases risk. Design professionals are not inventors. It is much less risky to stick with tried and proven materials, systems, and technologies. Design professionals provide a service. They do not produce a product that has been tested and perfected in a laboratory. A design professional's design work must work the first time.

Inquire as to whether the city or county (jurisdiction) in which the project is to be constructed has an easy-to-work-with building and planning department? Explore if the project will require review by a planning commission or architectural review board, and if a planning ordinance variance is required in order to construct it. Review the jurisdiction's history of reviewing projects before issuing building permits, and determine if it will likely take a long time. Sometimes the planning and review process can take months, even a year to navigate, and it can be expensive. Inquire as to whether the jurisdiction assigns plan-checking to a third-party reviewer. If so, this can add additional design expense, because third-party reviewers frequently have numerous review comments that must be answered, many of them designed to reduce their own risk as plan-checkers.

Study the project design budget and determine if it is sufficient. If the design team was involved in establishing the budget, it probably is. However, this doesn't always happen—sometimes business developers negotiate fees without input from a design team. Other times they get input from a team but negotiate and agree to a reduced fee in order to win the job. When this happens, the amount of money the project manager has to work with is less than originally expected. If the agreed-to design fee is too low, the project's potential financial performance may be at risk right from the beginning. The PM may have to be very clever in planning the work and very effective in performing it. In extreme cases, the PM may have to make the difficult decision to either tell upper management that the project can't be accomplished for the design fee or try to manage the project on budget, regardless of how difficult it may be. Both scenarios involve risk. Project managers play it both ways, with varying degrees of success. Some worry themselves sick because they were handed an impossible job. After trying it both ways myself, I now know how I play it: I tell upper management, rationalizing that we are all in this together, so let's share the risk.

Calculate whether the owner's construction budget is sufficient or not. Owner budgets are frequently based on outdated historical construction cost data. Sometimes they are based on studies done years earlier and never updated. Other times, the construction cost is politically determined. Boston's Big Dig is an example. In 1987, the project received $2.5 billion in federal funding. In 1991, the project was brought before the U.S. Senate again for an additional $2.5 billion in funding. Some skeptical senators wondered if the project might not return for a third time, but Massachusetts Senator Edward Kennedy told them, "There is no intention of repeating or coming back for additional resources."[(45)] As it turned out, the Big Dig returned to the well many times and wound up costing $16 billion. Had its proponents asked for $16 billion at the outset, it is very doubtful the project would have ever been built. In my experience, owners must be told at the outset if their budgets are unrealistic. They may not want to hear it, but they need to know it. If they are not told upfront, they will certainly learn it the hard way after they have spent their money on design. They are not likely to be happy when bids come in well over their unrealistic budget. Then it will be their turn to wonder what the project manager and design firm were thinking.

Question if the project schedule is realistic. Clients are always in a hurry. Many do not fully appreciate just how long it takes to flesh out a project program, consider design alternatives, select one with buy-in from all the owner's user-groups, and then produce construction documents. If the schedule is not realistic, the project is at risk. The client may be very disappointed when it's not completed on time. Like construction budgets, owners need to know if their project schedule expectations are unrealistic. If they find out at the end, when the project is months or even years behind schedule, they may blame the project manager and design firm.

Make sure there is a licensed individual for each discipline. All states require licensure in order to practice architecture and landscape architecture, civil, structural, mechanical, fire protection, and electrical engineering. Some states require interior designers to be licensed. In addition, states have different ownership requirements for design firms practicing within their state. It is important to understand the design professional licensing laws in the state where the project is to be constructed. For the project manager, this means that every discipline represented on the project must have a lead design professional licensed in the state where the project is to be built. If not, then some team members will have to attain their state licenses before the design of the project is completed.

Consider the dynamics of the team. Design team members that have worked together before know each other's capabilities, strengths, weaknesses, and idiosyncrasies. Because they do, projects tend to move along smoothly and coordination issues are frequently resolved with little effort. PMs may not know exactly what to expect from new team members. Are they qualified, will they perform on time, will they coordinate their work, will they make too many mistakes? New team members are unknown commodities and this increases project risk. It behooves PMs to watch new team members closely for a while, until the PM has confidence that they perform as expected and work well with the rest of the design team.

Prepared a project work plan. The project must be planned before it can be executed. Without a work plan, project risk increases. Preparation of a project work plan includes six basic steps:

1. Define the project scope of work in sufficient detail so the design team knows what is expected.
2. Identify the project team members and define who is responsible for which elements of work.
3. Prepare a project work breakdown structure (WBS), which is the process of dissecting the project into manageable tasks and assigning budgets and durations to each task.
4. Develop the project schedule.
5. Establish the requirements for the project quality control program.
6. Identify the various project-specific procedures and standards that all team members are expected to follow.

Work breakdown structures (WBS) must be in enough detail to permit effective management. A poorly formulated WBS increases project risk because it increases the difficulty of knowing whether the project is progressing as planned.

Starting a project without a documented work plan puts the project at great risk. Many design firms require PMs to prepare a written project work plan before project work begins. The project work plan document provides the basic project information and describes the planning activities and procedures necessary for team members to effectively do their work. My book *Project Management for Design Professionals* (Ramroth 2006) includes a suggested outline for a project work plan document. I have included it here because it serves as a good checklist for developing a comprehensive project work plan.

Outline of a Typical Project Work Plan Document

1. Project Objectives
    - General description of the project, including any unusual project features
    - Major project objectives
    - Major milestones (project phases)
    - Project construction budget, if known
    - Deliverables by milestones or phases: often takes the form of lists of drawings, specifications, reports, etc., required at the end of each phase
2. Project Team
    - Project organization chart
    - List of key team members, names, affiliations (in-house or consultants), phone numbers, and addresses
    - Roles and responsibilities of key team members
3. Project Task-Budget Information
    - List of project tasks and their budgets (expressed as dollars or hours)
    - WBS task codes (time card charge codes)
    - ODC (other direct cost) budgets (expressed as dollars)
4. Project Schedule Information
    - Project schedule
    - Project milestone information
5. Project Quality Control Procedures
    - Quality control plan
    - Identification of quality control manager
    - Quality control procedures
6. Miscellaneous Procedures
    - Accounting/Invoicing information
    - Project safety procedures
    - CADD and drafting standards and procedures including identification of CADD manager
    - Project filing procedures
    - Telephone conversation and meeting minutes standard formats

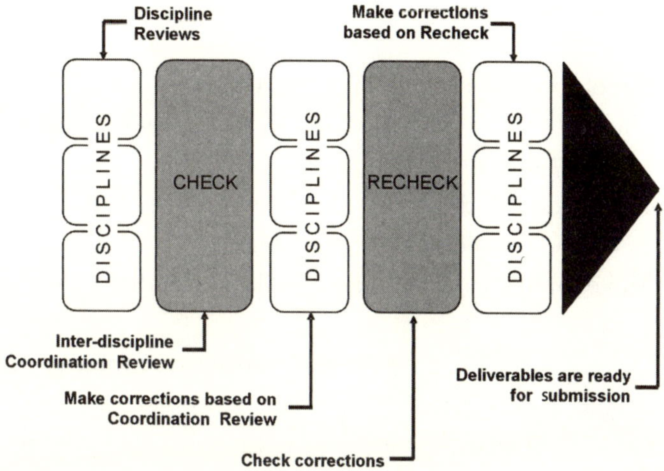

FIGURE 5.3 *Quality control review process flow diagram.*

The PM should distribute the project work plan document to all team members, the design firm's accounting department, and appropriate management personnel. It is a good idea to go through the document with the design team in a project kick-off meeting. Team members may have questions or concerns that the PM can answer. As the project moves along, some aspects of it are likely to change. When this happens, the project work plan should be updated to reflect the changes. If this is not done, the document eventually becomes out-of-date and of limited use.

MANAGING RISK WHILE EXECUTING THE PROJECT

Once the project starts, there are a number of preventive risk management activities that the PM must perform. Without them, project risk increases. The following list summarizes the major ones.

- Inform the owner of potential problems.
- Keep the owner informed of decisions and involved in the project.
- Develop a good working relationship with the owner. Happy owners are much less likely to find fault when minor things go wrong.
- Document all decisions and keep good records.
- Keep the project moving on schedule.
- Make timely decisions and make decisions in the proper order.
- Do not start production of contract documents until the design is well defined.
- Get the client's approval in writing before moving to the next phase of work. Some agreements require the owner to approve the design phase work and issue a notice-to-proceed before the production of contract documents begins. The provision helps protect both the design firm and the owner from untimely and costly changes to the work during the contract document phase.
- Properly coordinate interdisciplinary projects.
- Check all work products before they leave the office.

Every project should be checked following the design firm's standard quality assurance and control process. Figure 5.3 on page 141 shows a typical quality control process flow diagram.

Many owners believe that construction drawings and specifications should be perfect. A commonly used phrase in business development literature goes something like this, "We will check our work to ensure that it is accurate." This is tantamount to saying, "We will guarantee that the work is perfect," and that is impossible. To reduce risk, design professionals should not lead owners to believe that they will produce a perfect set of construction documents. Instead, design professionals should educate owners so that they understand that construction documents are rarely if ever complete and free of errors.

It is important to design to the construction budget. Owners want to build their projects. They are not interested in paper projects that are too expensive to be built. For many owners, projects that wind up costing double their budgets are useless. Budgets must be realistic and the design executed with respect for the project budget. There is more about designing to budget in Chapter 6.

Check the project's financial performance regularly. Design firms need to make sure that project managers receive timely financial information regarding their projects. If PMs don't know where their projects stand financially, it is difficult to keep the projects on budget. In addition, firms need to know that PMs are looking at the financial information and managing their projects' budgets.

To this end, design firms should conduct project reviews on a regular basis. Larger firms have so many projects that it may not be possible to review all projects regularly. Two approaches can be used to resolve the time issue. One approach is to review only projects above a certain dollar-sized fee. The amount will vary from firm to firm. The 80/20 rule can help decide which projects to review. Review the largest 20 percent of the projects, because this is likely to result in checking the management of 80 percent of the total design work in the office. Another method is to review only those projects that have already indicated problems. The disadvantage to this method is that it only singles out poor-performing projects. The firm does not get much feedback or insight from successful PMs with successful projects. Nor can the review identify potential problems and perhaps offer suggestions for heading them off. It also means that if a PM has to attend a project review, everyone else in the office knows that this PM has a problem.

My experience is that it is wiser to review both good and bad projects. Reviews can be private or held in the presence of other PMs. One office manager that I worked for held project reviews with all PMs in attendance. The reviews were monthly, and he chose at random a few projects to review. Of course, it was not completely random. He chose projects that had problems and some that were performing beautifully or had recent successes, requiring PMs to air their problems, but also giving them opportunities to talk about their successes. This had the affect of equalizing the playing field for all PMs. The reviews were handled matter-of-factly and were intended to be instructive, not to tear down PMs.

Project managers should avoid performing design work that is not a part of the scope of work. Perform what needs to be performed and no more. If two schematic alternatives are requested, don't do three. Many designers do not know when to stop designing. Given enough time, all good designers can probably come up with two or three more possible solutions. With many projects, however, that is a luxury the design firm cannot afford—and that the owner won't fund. I am aware of a project manager that had a perspective prepared for a client's project with the idea of giving it to the client as a "gift." The client was very appreciative but quickly wanted this changed and that changed and something added. The PM agreed to make the changes, which were costly even though the changes weren't meant to be in the scope of work.

CHAPTER SUMMARY

The project manager is essential to the success of a project. Project risk increases when project managers do not have the right combination of skills. Good project managers have the following characteristics:

- Desire to be a project manager
- Organizational skills
- Design and technical competency in the project type
- Generalist's attitude
- Ability to make decisions with incomplete information

- Professional and ethical behavior
- Strong communication skills
- Ability to motivate design teams
- Risk management capabilities

Although projects fail in many ways, there are six fundamental reasons.

1. The project was impossible to begin with due to technological and/or knowledge limitations.
2. One or more project objectives were unrealistic and therefore impossible to achieve.
3. There were reasons beyond everyone's control, such as acts of nature.
4. It was impossible to succeed because of the way the project was sold.
5. Execution of the project was poorly planned.
6. Project work was poorly executed.

Project risk management needs to start before the project becomes a project. Before it starts, the following project risk management steps should be taken:

- Make sure that a project manager with the right skill set is assigned to the project. Also make sure that an appropriate project designer is available.
- Have the project manager and design team participate in establishing the fee.
- Make sure there is staff available to perform the work.
- Examine the project for unusual scope, budget, and schedule requirements.
- Analyze the project's proposed construction contracting method.
- Identify the project's politically sensitive issues.
- Sell comprehensive services whenever possible.

Project risk decreases when projects are planned before they are executed. There are six objectives of the project work plan:

1. Define the project scope of work
2. Identify the project team
3. Breakdown the project into manageable tasks and assign task budgets
4. Develop the project schedule
5. Establish the project quality control program
6. Identify project-specific procedures and standards

Product managers should consider the following before embarking on the execution of a project:

- Determine if the project can be done with all in-house staff or if consultants are needed.
- Verify that the project design budget is sufficient.
- Make sure that a contingency fund has been set aside.
- Do not attempt to run projects without a contingency.
- Verify that the owner's construction budget is sufficient.
- Determine if the owner's requirements are fixed or subject to change.
- Execute the project using tried and proven materials and systems.
- Determine the governing jurisdiction's (city or county's) planning and building requirements and understand their requirements, time schedule for review, and any planning commission or review board procedures.
- Determine that key discipline leads have the necessary state licenses.

There are a number of things that can reduce risk while executing the project, including:

- Keep the client informed of decisions and potential problems.
- Avoid performing work that is not part of the scope of work.
- Conduct quality control checks of documents prior to every submittal.
- Document decisions and keep good project records.
- Make timely decisions—keep the project moving on schedule.
- Design to construction budget.
- Do not start production of the contract documents until the design is well defined and the owner has approved the preliminary design.
- Remember that design is iterative while the production of construction documents is linear.
- Design firms should have regular project reviews to determine the status of projects and their financial performance.

Chapter

6

COST, CONSULTANTS, AND CODES

It is a hectic and anxious time. Just prior to 2:00 P.M. on a Tuesday, Wednesday, or a Thursday—rarely a Monday or Friday—contractors mill around in a building lobby or anxiously pace just outside. Over their cell phones the bid runners receive last-minute bids from subcontractors and instructions from the home office. They write them down and frantically assess the situation trying to determine who their competitors are and how many bidders there are. They are hoping there aren't many because competition drives the bids down.

Among the contractors there is frequently an anxious design professional. He or she is also assessing the situation, wondering the same things, but thinking "the more the merrier," because, like the contractors, the design professional knows that the more bidders there are, the greater the likelihood that bids will be low. I recall a bid opening where a contractor had two envelopes in his jacket pocket. After he determined that he was the only bidder, he chose which envelope to submit—the higher of the two, no doubt.

The design professional also wonders how the bids will compare to his or her estimate. The owner's project manager is fretting about the bids, too, worrying if the lowest bid will be within budget. Both glance frequently at their watches. In a half-hour or so, the owner and the design professional will know the answer, and it will be over.

Minutes before 2:00, the contractors finish their bids, seal their envelopes, and give them to the owner's bid-representative. The owner's representative time-stamps the envelopes. The contractors take their seats. Occasionally, a bidder hands in a bid too late. The owner's representative hands the envelope back apologetically. With a harsh exchange of words or tears, the bid runner stomps out. However, if the bid runner is smart, he or she will come back and take notes as the bidders and amounts are read.

Shortly after 2:00, the bids are opened. One at a time, the names of the contractors are announced and their bids read aloud. The contractors write down the bids. The owner's representative gives a cursory look to see if all the supporting paperwork is included: the bid bond, the performance bond, and so on. When all the bids are read, the owner's representative thanks everyone and may or may not announce who is the apparent low bidder. Exactly who will ultimately be the low bidder depends upon whether all the supporting paper work is in order and that will take days or sometimes weeks to determine. Nevertheless, the contractors know who the apparent low bidder is. That bidder leaves happy, and the others leave disappointed. The owner and the design professional are either pleased or disappointed. They now know: either the project is within budget or it isn't. It was a good day or a bad one.

DESIGNING TO CONSTRUCTION BUDGET

Designing a project that does not exceed its construction budget is not easy and it does not happen by accident. It requires a reconciliation of owner expectations with the owner's pocketbook. It requires a squaring of project scope with the costs of building materials and systems. It demands a cost-conscious approach to design, selection of materials, systems, and detailing. It takes continuous monitoring of the design process to keep both project scope and construction costs from slowly increasing. If any of these are allowed to run unrestrained, a gap is likely to form and widen between the project's budget and its design. If left unattended, it may widen into a deep chasm. If this happens, it is almost certain that the project will blow its budget.

There are many projects in history that have gone seriously over budget. As discussed earlier, Australia's Sydney Opera House, the Scottish Parliament Building in Edinburgh, and Boston's Big Dig are three notorious examples. In the long run, project cost overruns are frequently forgotten. The public no longer cares about the original cost of the Sydney Opera House. The building has since become an icon for the city, and indeed, for the entire country of Australia. The same is likely to be true for the Scottish Parliament Building and Boston's new skyline. Eventually, both will add considerably to the fame and economies of their respective cities. Nevertheless, at the time, these three projects caused considerable consternation and were the subject of frequent rants from critics.

The same can be said for the countless lesser-known projects that exceeded their construction cost budgets. Design professionals' reputations have been tarnished, temporarily or permanently, due to cost overruns. Many owners have sued their design professional when a project grossly exceeded its construction budget. I recall a film-processing laboratory that went considerably over budget. The low bid was double the design professional's construction cost estimate. As far as the owner was concerned, a film-processing facility that cost twice what it was supposed to was useless. Very irritated, the owner stopped paying the design firm and hired an attorney. The design firm spent considerable time and money defending itself and trying to receive payment on its final invoice. Eventually a compromise was reached and the design firm received partial payment on its outstanding invoice, but not enough to cover the cost of attorney fees.

For design professionals, letting projects consistently run over budget is not good business. For many owners, a project that was supposed to cost *X* but costs 2*X* has no value. The economic calculations that once made it viable just don't apply. Consequently, many owners look for cost-conscious design professionals for their projects. This doesn't mean they are looking for cheap designs; it means they are looking for design professionals that control the costs of their designs. When seeking statements of qualifications (SOQs) from design professionals, owners frequently ask for data regarding past project performance. They ask for references and information regarding the design professional's estimates compared to actual construction costs.

Projects on budget are far less risky than the alternative. Remaining on budget requires a cost-control plan established and implemented from the outset of the project. The major steps of a construction cost-control plan include:

1. Determine the project's scope and budget
2. Break the budget down into smaller manageable pieces
3. Build in a contingency
4. Design to scope and budget
5. Control scope and cost creep
6. Monitor progress regularly
7. Reconcile cost increases
8. Warn the client of cost effects beyond the design professional's control

Determine the Project's Scope and Budget

The first step is to determine if the owner's budget is realistic. Many owners have an idea of what they want to build and how much they want to spend to build it. The problem is that the two don't always mesh. The design professional needs to determine if the owner has realistic expectations. If not, the design professional needs to tell the owner at the outset. If there is a big disconnect between the owner's project and pocketbook, there may be nothing the design professional can do through design to resolve it. Telling the owner upfront about unrealistic expectations is the best way of controlling the risk of cost overruns. The owner can reduce the scope of the project, add money to the budget, or do a bit of both.

In some instances the owner has a budget, but needs assistance from the design professional in determining a realistic scope of work. Still other times, the owner asks for assistance with both determining a scope of work and a budget. In all cases, the first step in project construction cost control planning is to develop a scope of work and a budget that are in sync with each other.

Break the Budget Down into Smaller Manageable Pieces

There are various ways of breaking down a project. A common way is to establish separate budgets for all the major components or materials and systems. For example, in a building project the major components might be site work, foundation, structural steel framing, exterior cladding, roofing, finish materials, and mechanical and electrical systems. Separate budgets can be established for each major component. The budgets for each component are shared with all disciplines. Each discipline—civil, architecture, structural, mechanical, and electrical—is responsible for designing within its budget. An alternative is to break the project down by CSI (construction specifications institute) divisions. This has the advantage of setting up budgets that will eventually track well with material and system specification sections. Because detailed estimates of probable construction cost are frequently formatted by CSI specification sections, it is easier to check and compare costs against budgets because budget costs and specification sections coincide.

Build in a Contingency

It is impossible to break down a project into smaller budgets perfectly. Sometimes the design of certain systems or the costs of some materials winds up costing more than budgeted. In an ideal world, just as many areas would run under budget as over, but this rarely happens. Consequently, it is necessary to have a contingency. The contingency is frequently expressed as a percentage, which varies depending upon the project. Because the contingency is established at the outset—when there is the greatest number of unknowns—the contingency percentage needs to be relatively high. Contingencies of 20 percent, 25 percent, and even 30 percent are not uncommon. As the project progresses and more information becomes known, portions of the contingency can be moved to bolster budgets in specific areas. Without a contingency the only way to increase a budget in one area is to reduce the budget in another—rob Peter to pay Paul. The only other alternative is to increase the overall project budget, which is often an untenable solution. In any event, it is not a solution that the project manager can make unilaterally, the owner needs to participate.

Control Scope and Cost Creep

They say a frog will stay in a pan of water as the temperature slowly rises. He doesn't jump until finally it is too late. Scope creep is the slow, incremental increase in the amount of work. It adversely affects both design costs and construction costs. Adding a door here, skylight there, closet here, and sink there has cost consequences. String enough small changes together and it can add up to a big problem. To control scope creep, the design professional needs to keep track of how the changes affect the overall construction budget. The budget contingency—if it was a healthy one—can possibly absorb a few changes, but there is a limit. Design professionals can't be like the frog in the pan and do nothing until it is too late. They have to let the owner know of the cost implications of changes as they occur.

In addition to owner changes, cost creep can occur due to design. The design disciplines can increase construction costs if they design with materials and systems that are more expensive than others are. Project team members need to be cognizant of the effects that their choices of materials and systems have on construction costs.

Monitor Progress Regularly

Once the budgets are set and design is progressing, the process has to be monitored. Scope and costs can increase at almost at any time during the design process. There are a few key points during design when it is a good idea to give the project a cost-sanity check. A check needs to occur at the end of schematic design and at the end of design development. By the end of design development, the design should have enough detail to identify all major materials and systems, making it possible to add significant detail to the construction cost estimate prepared at the end of schematic design. After the design development, the project goes into production of construction documents. It is much more costly to make changes during the construction document phase than during early project phases. Another good time to check is just prior to bidding or construction. The project is then fully designed and a detailed cost estimate can be developed.

Reconcile Cost Increases

After each estimate, the scope and costs have to be reconciled. If the project is too expensive at the conclusion of schematic design, the situation will not improve after design development. The budget problems must be fixed as the project moves along and this often requires cutting scope. Waiting to reconcile cost and scope differences until the final estimate just before bidding leaves you no other option but to identify portions of the work as "add alternates." This lowers the cost of the base bid, but it also means that some portions of the work may not be built due to cost restraints.

Warn Client of Cost Effects Beyond the Design Professional's Control

The owner must understand what the project budget means. Without a definition, owners may jump to the conclusion that the project budget is "set in stone" and something like a guaranteed maximum price. It is no such thing; it is an estimate only, a probable cost, not a prediction of the actual cost. Many design professionals no longer use the term "construction cost estimate." They use the term "estimate of probable construction cost" to convey better the message—and warning—that the actual construction cost may vary from the estimate.

To control the risk of negative fallout that may occur should bids be higher than expected, design professionals must take a number of precautions. To begin with, the wording of the agreement is important. It should not contain clauses such as, "the design professional warrants that the cost of the project," or, "the construction budget represents the project's guaranteed maximum price," or "the design professional ensures that the project budget will not be exceeded." Better alternatives include wording such as: "The design professional will endeavor to develop a budget for the project that attempts to define the costs of the project. The owner recognizes that the project budget does not represent the actual dollar amount of the bids received." Alternative wording is: "The project budget represents the design professional's best judgment of the project costs. The owner recognizes that the design professional has no control over the bidding climate or the manner in which contractors develop their bids, both of which affect the project costs." The design professional has no control over how the owner advertises for bidders, how many bidders there will be, how busy contractors are, the cost of building materials, and so on. In 2005 and 2006, the costs of some building materials skyrocketed, steel and concrete being two good examples with costs of both materials experiencing yearly double-digit percentage increases.

CONTROLLING DESIGN BUDGETS

Some of the components of construction cost control are also applicable to design cost control. Controlling design costs requires a plan. It requires determination of the project scope and reconciliation of it with the design budget. As discussed in Chapter 5, the design budget must be broken down into manageable tasks and a contingency established. The time spent designing must be consistent with the budget limitations. Project managers must guard against scope creep so the design team doesn't spend time designing things that are not within the original scope of work. Progress must be monitored on a regular basis. If there are signs that portions of the work are going over budget, the project manager has to make adjustments. Ignored problems usually don't go away—they just get worse.

CONSULTANTS

It is no secret that architects and engineers frequently take on projects that they cannot fully perform with their own staff. Consultants are needed to round out the design team. When a design professional agrees to provide services for a client, the level of responsibility is the same whether the work is performed by in-house staff or by consultants. Consequently, choosing the right consultants is important. Often it is crucial to a project's success. Risk is best managed when consultants have the same interests in risk management as the design firm.

Design firms occasionally select consultants because the consultant is an "owner favorite." On other occasions, the consultant is selected because the firm meets a DBE (disadvantaged business enterprise) participation goal or small business set-aside requirement that the owner wishes to achieve. Choosing consultants only for these reasons can significantly increase project risk.

Choosing new consultants is much like hiring a new employee. The PM should choose consultants that are qualified and that strengthen the design team. The consultant's performance will affect the project and reflect on the design team and design firm, so choose carefully.

Before selecting new consultants, they should be interviewed. It is good to meet with them and discuss their qualifications and past work experience. It is worth asking for references. Visiting their office is helpful, because it allows the PM to meet the staff members who are likely to be assigned to the project. PMs should confirm that the consultant has the necessary professional licenses and can meet the project's insurance requirements. Most importantly, the PM must decide if the consultant will be a good fit with other design team members.

Sometimes it is better to use consultants than in-house design professionals. There are three main reasons:

1. The consultant has a good working relationship with the owner and the design firm wishes to take advantage of the relationship.
2. The consultant has better or specialized experience in a particular building type or service.
3. The consultant is less expensive than using in-house staff to accomplish the same work.

I recall many instances when a consultant strengthened our in-house design team's qualifications because it had worked for an owner and had already forged a good relationship. Under these circumstances, using a consultant for work that could be performed in-house can increase the likelihood of being selected. Other times consultants have experience with certain project-types that are lacking from in-house disciplines. For example, a multidisciplinary design firm may have structural engineers, but limited experience designing school projects, high-rise office buildings, power plants, or wastewater treatment plants. The same may be true for mechanical and electrical engineering. Frequently large E/A firms lack experience in architecture for certain building types. In these cases, it may be less of a risk to use a consulting architectural firm that has the expertise that is missing or sketchy from among in-house staff.

Consultants are sometimes less expensive than using in-house staff. In-house staff members are marked up by a labor multiplier. The multiplier frequently hovers around 3. This means that if a staff member's salary is $40 per hour, the owner is invoiced and the job cost accounting system bills the job at $120 per hour. A consultant may charge only $90 per hour for the same work, which is a savings of $30 per hour. On lump-sum projects, the savings can add up and have a significant positive effect on the job's bottom line.

Consultants do not have to be from the same geographic area as the design firm. Some areas of the country have higher costs of living than others do and the labor pool reflects this. On many occasions, I have worked on projects where consultants have been from a different area of the United States where salaries were lower. My experience has been that the quality of work is comparable, regardless of location. However, it is necessary to confirm that the consultant is licensed in the state where the project is to be built. It also may increase the coordination work for the PM, but generally this is no more time consuming than using a consultant from around the block.

CODES

The purpose of building codes is to control the risk of something going wrong that could adversely affect public safety. Laws regarding the safety of buildings have been around for millennia. The ancient Roman historian Cornelius Tacitus (c. 56–120) wrote about a change in ancient Rome's building laws in his *Annals of Imperial Rome.* In 27 A.D., a cheap and shoddily constructed amphitheater suddenly collapsed, injuring and killing 50,000 spectators. Rome's citizens were outraged and demanded safer buildings. The Senate quickly banished the amphitheater's builder and passed a law that set a minimum net-worth for those who undertake the design and construction of large public building projects. In addition, the Senate established structural requirements for public projects and enforced procedures that required official inspections of the foundations and super-structures of public buildings under construction.[46]

Over the course of centuries and as a result of many disasters, countless improvements have been made to building codes and planning ordinances. Beginning with New York City's Zone Regulations of 1916, building code and planning requirements were separated into two different types of documents: building codes and planning ordinances. Building codes address how to design and construct buildings. Planning ordinances address how the building fits within the greater fabric of the city, county, or other jurisdiction. Planning ordinances address building use, setbacks from streets, lot coverage, height, off-street parking requirements, landscaping requirements, and so on.

Design professionals cannot disregard code and planning requirements. They establish the ground rules for design and construction and they have a significant impact on the design. The Chrysler Building and the Empire State Building are stepped skyscrapers because New York City's zoning regulations require tall buildings to step back away from property lines as they ascend skyward. If code and planning requirements are not addressed and followed, then the design will never make it from two-dimensional paper to the three-dimensional real world.

Building designs are submitted to local building and planning departments for review. Building plan checkers, planning department staff, and fire marshals review the drawings to make sure that the proposed building complies with the scores of code and planning requirements that bear on its design. If the project complies, a building permit is issued. If it does not, a list of deficiencies is issued instead. Until they are addressed and resolved to the satisfaction of the building department, construction cannot begin.

In addition to building and planning regulations many other regulatory agencies may have requirements that affect a project. There may be OSHA regulations to abide by, health department requirements, air-quality management district requirements, public works requirements, coastal or river commission regulations, and so on.

At the outset of the project, it is good to identify all the regulatory agencies that may have jurisdiction over the project and determine the requirements that will affect the design of the project. Before the design gets too far along, it is wise to check that it complies with all the requirements. If this is not done, the project manager, project designers, owner, and design firm may be unpleasantly surprised when the code analysis is finally done just before submission to the building department, and reveals numerous deficiencies. At that point, there may be a number of changes required in order to make the design comply. Projects near completion are much more expensive to correct than those in the beginning stages. A worse scenario is when a design firm does no code and planning review and the project is submitted to the building department for plan check. Instead of a building permit, the building department issues a long list of deficiencies that must be addressed and corrected. If and when this happens, owners may wonder about the competency of their design professionals.

Consequently, design firms should include code analysis procedures as part of their comprehensive risk management plan. A code analysis plan should

- require all project managers to include code analyses in the project work plan;
- establish a list of qualified individuals within the firm to conduct code reviews;
- encourage or require project managers, when the project warrants it, to schedule and attend preliminary review meetings with planning and building departments; and
- set up a standardized code analysis process—a code review format—because a format eliminates the need to re-determine one for every project.

The code analysis form can take a variety of formats such as table, list, or paragraph form. The format is a matter of preference. However, it must contain the information outlined in the next section. At the beginning of the project, some of the information about the specific design will be unknown, but much of the code and planning information can be determined before the design begins. For those items associated directly with the design, it is generally possible to make a few educated guesses or assumptions that are likely to be correct. For example, the building type (I, II, III, IV, or V) is frequently known before the design begins or known very soon thereafter. In addition, the occupancy group or groups are generally known from an educated reading of the project program of spaces or functions. For the first go around the information is used for design. At the end of the project, the information is used to confirm that the final design meets code and planning requirements and that the information is logically presented for the building department's review.

A code analysis is not something that should be done just before the project goes to the building department. It should occur at or near the beginning of the project. Knowing what codes and ordinances apply and knowing their requirements are essential basic programming requirements. They are as important as the owner's program of spaces. Together they drive the building design.

Code Analyses

Code analyses help designers understand the code implications that drive design decisions and building detailing. For example, large assembly spaces must satisfy code requirements that address the number of exit doors, corridor widths, flame spreads of finish materials, as well as other requirements. Fire walls may be required between different occupancy types. If there are hazardous materials in the building or portions of it, there are requirements for their proper storage and containment should a spill or leak occur.

Project managers should assign someone to be responsible for code and zoning review. On small projects this may be the project manager. On larger projects the review is frequently more complicated, requiring someone with the proper expertise to conduct the review. The PM may or may not have these qualifications.

A building and planning code analysis should address and/or include the following information:

- All codes governing design and construction
- Planning and city ordinance requirements
- Other jurisdiction regulations
- Fire marshal requirements
- Special governmental agencies, districts, commissions, or associations requirements
- Area summary calculations
- Height summary information
- Accessibility requirements

- Identification of hazardous materials
- Egress or exiting requirements
- Special owner requirements

Codes Governing Design and Construction. These include any energy efficiency codes or requirements. Be sure to reference the name of the building code, the edition date, and whether the state or city/county has any amendments to the code.

Planning and City Ordinance Requirements. This includes identification of zoning districts, setback requirements, allowable lot coverage, off-street parking, landscaping requirements, height restrictions, sightlines, art requirements, and/or building shadow requirements. It should also address whether there are any special architectural style requirements and whether or not an architectural review process is required, and what those requirements are.

Other Jurisdiction Regulations. These include OSHA, air quality districts, city health departments, and so on.

Fire Marshal Requirements. It is becoming more common for cities to require fire suppression systems for buildings over a certain size, such as 5,000 square feet, for example.

Special Governmental Agencies, Districts, Commissions, or Associations Requirements. For example, there may be air-quality management district requirements, a coastal commission, an environmental quality act, or commission requirements. There may also be storm water and sanitary sewer discharge requirements.

Area Summary Calculations. This includes both actual and allowable calculations based on construction type and occupancy classification. This calculation determines if minimum setback distances from property lines, right-of-ways, or other buildings sharing the same site are required in order to increase a building's allowable area. It also determines whether an automatic fire suppression system is required to increase the building's allowable area.

Height Summary Information. This includes actual height and allowable height based on construction type and occupancy classification.

Accessibility Requirements. These are such things as identification of accessible parking spaces, building entrances, accessible toilet facilities, signage, and so on.

Identification of Hazardous Materials. If the project includes the storage or handling of hazardous materials there are spill containment, fire suppression, exit devices, detection, and alarm system requirements that must be met. There may also be fire-rated occupancy separation requirements.

Egress or Exiting Requirements. These are based on use and number of occupants.

Special Owner Requirements. A common requirement that is growing in popularity is the U.S. Green Building Council's Leadership in Energy and Environmental Design (LEED) requirements. Owners may wish to achieve a specific level of LEED certification for their project—silver, gold, or platinum. Although not a code requirement, meeting the various conditions it sets for LEED certification requires an almost code-like approach.

Architectural Review Boards

The discussion of architectural style requirements reminds me of a small project I worked on many years ago. The project was located in a small tourist town that had some very specific requirements regarding the "look" of all buildings located in its central business district. All buildings had to look Bavarian, as if from the German/Swiss alpine region from the mid to late 19th century. Specific criteria were established for the use of materials and the "look" of all buildings—rectangular-shaped one, two, or three-story wood-sheathed or half-timbered-looking buildings with low-pitched roofs with expansive overhangs, elaborately decorated fascias, trim, finials, and wall murals. Any designer who ignored the city's requirements would be sadly disappointed when the architectural review board reviewed the project. Knowing and addressing the city's specific requirements was essential for a project's success.

Many communities have architectural review boards. After the 1925 earthquake, the City of Santa Barbara, CA, established criteria for appearance of its buildings. Buildings in Santa Barbara are reviewed by an architectural review board, which is interested in maintaining the city's California Mission-style character.

At some point after the schematic design phase but before the design development phase is completed, it is a good idea to schedule a preliminary review meeting with the building department. Building departments generally welcome this. It is helpful if a city planner and fire marshal attend the meeting and in many jurisdictions the building official will assist in scheduling their attendance.

During the meeting, the design professional explains the project and its responses to various high-level code and planning considerations. For example, a preliminary site plan can show setbacks, lot coverage, off-street parking and accessible access, location of fire hydrants, fire department access to both the site and building, and other pertinent major site features. The preliminary floor plan(s) can show the overall building area, area per floor, exiting pathways, location of exit stairs, elevators, areas of refuge, and so on. Although building officials will generally not give final approval to the plans during this meeting, they will ask questions, give advice, and warn the design professional of obvious violations or areas of concern. If the building or planning department officials and/or fire marshal have concerns, they will voice them in the meeting and the design professional can work on resolving the issues.

Resolving issues early-on in a project is much less costly than doing so after the contract drawings are complete. If all goes smoothly the design professional can have a reasonable degree of confidence that if the design proceeds along its current path, it will sail smoothly through planning and building department review when it is completed. It is a good idea to document the meeting with the building department officials and to send copies of the meeting minutes to all attendees.

Owners often wish to attend the preliminary building department review meeting. It is wise to discuss the meeting with the owner before setting it up. Some owners resent design professionals talking with building department officials without an owner's representative present.

Code Summaries

I recall a project where the project manager took the finished contract drawings to the building department for review. The project was simple and the design fee was small. The headache that followed was anything but. The project was a small addition to a much larger exiting building. The larger building was part of a complex of buildings that shared the same site—one building among processing buildings, warehouses, employee facilities, and administration offices, totaling several hundred thousand square feet. The buildings varied in height between one and three stories. They were built over a period of decades, and some were fire sprinklered and some were not. There had been little additions and remodels built here and there without permits. The project manager was busy with many other projects and since the project and design fee were small, an early code review was not done. The project went out to bid and into the building department at the same time. After about two weeks, the building department finished its review with a few fundamental review comments. The building official requested that an overall site plan be submitted, showing all the buildings and a table listing their areas, building types, and occupancies. In addition, it requested that a code analysis be done to show that the addition, which was only a thousand square feet or so, complied with the code requirements for allowable building area.

It only took a couple of hours to conclude that the addition, as designed, would not comply unless fire sprinklers were added to the new building and every other building without them. The 1,000-square-foot project, with a design fee of $20,000 or so, was now a monster. It would have been much better if the project manager had figured out the code implications of the addition at the outset of the project instead of at the end. It would have saved a lot of time and grief.

It is important that all major code and planning issues be addressed and resolved before the project transitions from iterative preliminary design to linear production. Code-related changes made during the production of contract documents can be costly and, depending upon the change, affect the work of many disciplines. Near completion, the project should be checked one final time to make sure that any changes to the design have not adversely affected code compliance.

At this point, it is a good idea to prepare a code summary sheet for the contract drawings. Code summaries are helpful to building officials and consequently make the review process go more smoothly. Many building departments require code summaries. It includes a summary of planning and building department requirements that bear directly on the design and the project's response to these requirements. The code summary generally includes a site plan and building plan or plans. The site plan shows setbacks from property lines, access to the site, the building footprint, lot coverage, number of parking stalls, and landscaping. The floor plan(s) show occupancy classifications, corridors, stair towers, fire wall locations, exit pathways, exit doors, and so on.

Building Development Review

One final comment regarding building code compliance: it is worth determining if the building department will review the drawings in-house with its own staff or by using a consultant hired by the building department. For many small and moderately sized cities and counties, it is becoming more common for projects to be reviewed by third-party design professionals who specialize in code review services. It is good to know this as soon as possible. It has been my experience that drawings reviewed by third parties almost never are approved the first time through.

In addition, plan-check services prefer to issue letters explaining their review comments, rather than marking the drawings with the necessary corrections. The review comments are often generated by editing a master list of dozens of possible review comments. Consequently, many review comments are general in nature and frequently there are many comments that don't even apply to the project. Nevertheless, third-party reviewers expect answers to all their review comments, relevant or otherwise. The process can be time consuming and expensive. To avoid the risk of project schedule slippage and spending a lot more money, it is a good idea to find out as soon as possible what the review procedure is likely to be and budget accordingly.

CHAPTER SUMMARY

Design professionals have two budgets they need to manage: the construction cost and the cost of doing the design work. Design professionals need to educate owners regarding the many factors that affect the cost of construction that are beyond the design professional's control, such as:

- How the project is advertised
- Whether the project is bid or sole source
- Bidding climate—how busy the contractors are
- How many bidders bid the project
- The escalating cost of materials and labor

The design professional must control the cost of doing the design work. To control the design budget, it is necessary to pay attention to scope creep, determine which discipline(s) are critical, and monitor design progress regularly. This can be accomplished by:

- Breaking the project budget down into smaller manageable pieces
- Building in a budget contingency
- Designing to scope and budget
- Controlling scope and cost creep
- Monitoring progress regularly, comparing the design to the budget
- Reconciling cost increases

Consultants must be selected carefully. Consider the following factors:

- Is the consultant qualified to perform the assigned tasks?
- Does the consultant strengthen the design team?

- Has the design firm worked with the consultant before and if so, was it a rewarding experience?
- Be wary of owner-recommended consultants. Interview them and check their references.
- Can the consultant comply with the owner and design firm's insurance requirements?

Consider using consultants instead of in-house staff when the

- in-house staff is too busy;
- consultant has a good working relationship with the owner;
- design firm can learn something from the consultant, expand its portfolio of work, and the design firm wishes to take advantage of the relationship;
- consultant has better or specialized experience in a particular building type or service; or
- consultant is less expensive than using in-house staff to accomplish the same work.

Building code review procedures and requirements should be part of a design firm's risk management plan. Design firms should:

- Require all project managers to include the requirement for code analyses in the project work plan.
- Establish a list of qualified individuals within the firm to conduct code reviews.
- Encourage or require project managers, when the project warrants it, to set up preliminary review meetings with planning and building departments.
- Set up a standardized code analysis process—a code review format—because a format does not have to be re-determined for every project.

- Determine if the reviewing jurisdiction reviews projects with its own staff or uses plan-check consultants, because responding to the latter's review comments can be both time consuming and expensive.

Chapter

7

PROFESSIONAL LIABILITY AND CONTRACTUAL RISKS

While studying architecture in college, I was assigned a space in the design studio to call my own. At most it measured 6 feet by 7 feet. My space had a drafting table with years of initials carved into it, a drafting stool, and a lockable base cabinet. I spent many nights and weekends in my space, working at my desk, missing out on other campus activities like dorm poker night, hanging out in the student union, and rock concerts. Nevertheless, I was not alone—there were plenty of fellow architectural students in the drafting studio with me—and I had the time of my life.

Because we spent so much time there, we personalized our spaces. Some students built make-shift partitions out of cardboard to which they taped their tracing-paper sketches. Some even made little roofs over their drafting tables and strung Christmas lights. Extension chords ran helter-skelter, powering our drafting lamps, electric erasers, and stereos.

Every month or so, the campus fire marshal would come through the studio, have a serious chat with us, and tear down some of the more egregious fire safety violations. We'd clean up our act for a while, but eventually go back to it, decking out our spaces better than ever.

PRODUCT LIABILITY VERSUS PROFESSIONAL LIABILITY

One of the appliances that the fire marshal hated more than any other was the popcorn popper. As every college student knows, the popcorn popper ranks as one of the most versatile gadgets ever invented. It pops popcorn, of course, but it can do so much more. It can heat up soup, chili, Spaghetti-Os, cook Top Ramen, and boil hot dogs—all perfect staples for hungry students pulling all-nighters.

Of course, cooking soup or anything other than popcorn, for that matter, is not a proper use of popcorn poppers and I am not advocating their use for such purposes. Here is the point: through numerous improvements, based on years of testing and real-world use, popcorn poppers have gotten better and better and have evolved into being wonderful, versatile appliances. In fact, the manufacturers of popcorn poppers are so good at making them that we expect every popcorn popper to work perfectly right out of the box every time. If it doesn't, we take it back to where we bought it and get another, no questions asked. If we use it and it starts a fire and causes injury (while popping popcorn, of course) we can hold the manufacturer liable for the damages it causes, because of a concept called product liability.

Product liability is the idea that manufacturers must produce products that are free from defects. If the product doesn't work properly or at all, or if it harms somebody while being used as intended, the manufacturer is responsible and must make amends for the defect or injury. The key to the concept of product liability is rather simple: anybody who is injured, financially or personally, by a defective product does not have to prove that the manufacturer was negligent. The law presumes that the manufacturer is *negligent.* It is up to the manufacturer to prove that it is not.

The same is true for builders. When a developer builds a tract house, the buyer can expect that the roof will not leak, even if the sales agreement says nothing about the roof. When an agreement is silent on an issue or the intentions are unclear, the law attempts to determine what the parties must have intended or assumed within the bounds of what is fair and reasonable.(47) In the case of roofs, it is fair and reasonable for a homebuyer to assume that the roof won't leak. Consequently, it is not necessary for the buyer to establish that the builder was negligent if the roof leaks. Rather it is clear from the outcome—the leaking roof—that the builder is at fault because the builder did not provide the buyer a house with a roof that does not leak.

Design professionals are not builders, nor are they like the manufacturers of popcorn poppers. There are millions of popcorn poppers that are exactly the same. Before the first one was sold, a lot of research and development went into perfecting it. This is not the case with the design of a building or other unique structure. Every building and structure is a bit different.

Design professionals provide services; they do not produce products. Because of this, product liability does not apply to design professionals. Another concept of liability does, called professional liability. Because the services that design professionals provide are not standard solutions to standard problems, design professionals are expected to exercise judgment and discretion based on their experience, expertise, and the specific interests and needs of their clients. Under these circumstances it is not possible to be perfect, nor is it expected. Instead, design professionals are expected to perform in a manner similar to competent professionals practicing under the same circumstances in the same areas of expertise, geographic location, and legal jurisdiction.

Under the concept of professional liability, if an owner is injured either physically or financially because of something the design professional allegedly did, the law does not automatically assume that the design professional is at fault. Instead, the law looks at the process that the design professional followed that led to the injurious outcome and compares the process to what other design professionals would have done under similar circumstances. University of California emeritus law professor Justin Sweet wrote that when a client hires a design professional, "American law has concluded that the *likely* understanding between the client and the professional designer is not that a *successful outcome* will be achieved when professional services are purchased but that the professional will perform as would other professionals."[(48)]

In general, a design professional is expected to:

- Accomplish the scope of work sought by the contract with the client.
- Do what is professionally reasonable under the circumstances.
- Perform as other design professionals would perform who practice in the same jurisdiction.

Note that the outcome is not the salient issue in determining fault—the process is. Although it is the undesirable outcome that prompts legal action, typical claims against design professionals revolve around the performance process. As previous chapters demonstrate, risk for design professionals increases if proper processes are not followed. Design professionals are judged relative to their peers and the processes they follow, therefore, if peers have taken steps to reduce the likelihood of undesirable outcomes, it behooves the design professional to implement similar procedures. It is like the saying, "When in Rome do as the Romans do." Because the law focuses on the process regarding matters of professional liability, the design professional is wise to focus on following sound processes, as well. Doing so reduces risk.

PROFESSIONAL LIABILITY INSURANCE

Some risks, should they occur, can result in significant losses to a design firm. A few can even jeopardize the firm's existence. One way to reduce these risks is to transfer them to someone else. As we have already discussed, transference is a common strategy used to control risk. The purpose of professional liability insurance is to transfer risk to an insurance company. In the traditional narrow view of managing risk, professional liability insurance is the main and sole component of risk management. In reality, however, it is more of a sharing of risks than an outright transfer. In exchange for paying professional liability insurance premiums, a design firm can expect the insurance company to absorb some of the costs of claims against the firm, but not all.

The purpose of professional liability insurance is to protect the design professional from claims due to so-called negligent acts. To prove that the design professional was negligent, a plaintiff must establish that:

- The design professional owed the plaintiff a duty to perform in accordance with some specific standard in order to protect the plaintiff from risk or harm.
- The design professional did not perform in accordance with the standard.
- There is a causal connection between the design professional's lack of performance to the standard and the plaintiff's physical and/or financial injury.[(49)]

From time to time, design professionals hear the term *errors and omissions* or *E&Os*. Many bristle at the terms and avoid using them as if they were swearing. The terms are little more than stand-ins for the term *negligence* as just defined. The terms *errors and omissions insurance* and *E&O insurance* are just other terms for professional liability insurance.

With most professional liability policies, there is a deductible. The deductible means that, up to a specified amount, the design firm must pay out-of-pocket for acts of negligence and attorneys' fees. The insurance company pays the amount above the deductible. Some firms are self-insured up to a certain dollar amount. For claims over that amount, liability insurance kicks in. Policies also contain upper monetary limits, meaning the policy will only cover costs up to a predetermined amount. Policies also have rules and procedures for preserving the coverage, meaning that if the rules and procedures are not followed the insurance carrier will not provide coverage. Consequently, design professionals must follow the stipulated rules in order to keep their coverage intact.

A basic professional liability policy provides more than just financial coverage in the event of a negligent act. The policy also provides legal defense. Most insurance companies have in-house attorneys. Although the attorneys work for the insurance company, their job is to represent the design firm. Some policies allow the design firm to select an attorney. Most policies are broad in nature and not only cover the firm but the firm's owners and employees, as well. It is important to confirm this with the insurance carrier, because some low-cost policies do not provide coverage for a firm's employees.

Insurance policies can be modified through exclusions and endorsements or a combination of both. If risks cannot be determined for a certain service provided by the design firm, this service can be excluded from the policy. If the design firm decides to venture into new areas of practice, such as design/build, coverage for these endeavors can be added by what are called endorsements.

In addition to general professional liability insurance, there is also project insurance. Project insurance is a separate insurance policy for a specific project that provides coverage for the project during design, construction, and beyond. The premiums and claims paid against project insurance have no effect on the firm's general professional liability policy. The duration of project insurance can vary, but 10 years is a common upper limit, meaning that the yearly premium is based on the understanding that it will be paid each year for a period of 10 years and coverage will be provided during the same time period.

Some firms choose not to carry liability insurance. Many sophisticated clients, however, require design firms to have professional liability insurance, and include specific insurance requirements in their owner/design professional agreements. If a design firm does not have liability insurance, it limits the types of clients and projects that the firm can pursue. For this reason, whether or not to carry liability insurance is a key factor in a design firm's overall business plan.

INDEPENDENT BROKERS AND INSURANCE COMPANY AGENTS

Design firms buy insurance two ways: through independent brokers and through insurance company agents. Insurance brokers are independent insurance professionals that represent their clients, not a specific insurance company, often called an insurer. Insurance company agents work directly for a specific insurer.

While independent brokers have the best interests of their design firm clients in mind, they may have limited access to insurance companies. Insurance company agents may only be authorized by insurers to sell certain types of policies and these, in turn, may be limited by territory or geographic area. An independent insurance broker can solicit various insurance companies on behalf of the design firm and receive quotes on premiums for various levels of coverage. The broker can provide professional advice regarding which combination of premium and policy coverage is best suited for the particular needs of the design firm. Insurance brokers and agents usually work on a commission, but they can also work on a fee basis.

Once a firm selects a policy, the broker or agent can continue to assist the design firm. They can review agreements before they are signed to see if they comply with the terms of the insurance policy. They can also provide constructive comments regarding a firm's organization and growth and how they might affect insurance. They can also help the firm in management techniques, not the least of which is managing the risk of claims.

Some questions to consider when choosing a broker or agent are:

- Does the insurance broker have access to many insurance carriers?
- Is the insurance company agent permitted to write policies for professional liability insurance in your geographic areas of practice?
- Can the broker or agent assist the firm regarding practice management techniques?
- Does the broker or agent seem qualified and have a commitment to help the firm?
- Is the broker or agent a person the design firm feels comfortable working with?
- Does the broker or agent work on a commission or a fee basis?
- Will the broker or agent be able and willing to provide service and advice to the firm for the commission or fee?

PROFESSIONAL LIABILITY POLICIES

Insurance policies should not be selected based on premium costs alone. There are a variety of factors that play into the insurance equation including:

- Cost and coverage of the insurance policy
- Experience of the insurance company in writing professional liability policies

- Ability of the insurance carrier to provide coverage in all geographical areas where the design firm does work
- Financial strength of the carrier

In addition, it is good to know if the insurance company is what is called an "admitted carrier." Admitted carriers comply with state rules for guaranteeing funds. If an admitted carrier fails, the state guarantees the coverage. Some carriers are not admitted carriers. Nonadmitted carriers are called excess-and-surplus carriers or E&S carriers. E&S carriers have greater flexibility in writing insurance and consequently can more aggressively price their policies. While this may have advantages in some cases, E&S carriers have the ability to move quickly in and out of insurance markets and geographic areas. If a nonadmitted insurance carrier decides to leave a geographic area or goes out of business, the state will not guarantee the insurance. Most state insurance commissions have a published list of admitted carriers and the policies that they write that are in compliance with state's insurance rules. Many state insurance commissions include the list on their websites.

Other factors to consider when selecting an insurance carrier include:

- Will the carrier be flexible in writing additional liability policies if needed? For example, will they write project insurance if needed and provide coverage for design/build ventures? Can the dollar amount of coverage be expanded?
- Does the policy cover the firm's employees as well as its owners? Some low-cost policies do not cover the negligent acts of employees when acting on behalf of the firm.
- What is the extent of exclusions? Insurance policies frequently include preclusions from coverage. Find out what they are and how they affect the firm's practice.

- What is the insurance company's claims-handling procedure? Does it handle claims in-house with its own attorneys? Will it let the design firm use its own attorney? When does it appoint a defense attorney?
- What is the level of the design professional's participation in handling claims? How much say does the design firm have in its defense?
- Is the insurance broker or agent able to assist the design professional in practice and risk management services? The range varies considerably. Some carriers provide extensive risk management services while others provide little or none.

OCCURRENCE, PRIOR ACTS, AND CLAIMS-MADE INSURANCE POLICIES

For design professionals, a claim may not occur until many years after the acts or services that caused, or allegedly caused, the claim. A faultily designed roof flashing detail, for example, may not leak for years. When it does, it could generate a claim. The design firm must have its professional liability insurance policy in effect in order to be covered. Consequently, design firms must have liability insurance coverage at all times, from the time the design work starts until long after the project is built.

Professional insurance liability policies fall into one of three policy-basis categories:

- Occurrence
- Prior acts
- Claims-made

An occurrence-basis policy provides insurance coverage even after the policy is cancelled, so long as the policy was in effect when the acts or services—or alleged acts or services—that caused the claim occurred. A prior acts policy covers acts and services that took place *before* the insured first became insured or was insured by a previous insurance company.

Claims-made basis policies are the most common type of professional liability insurance policies. They require that claims be made when the policy is in effect. Each year, by paying the premium, the insurance continues in effect. Of course, the cost of the premium may change—increase—from year to year.

The typical homeowner's insurance policy is an example of a claims-made basis policy. As long as the homeowner pays the yearly premium, the homeowner is covered. If there is a storm and a tree falls and crushes a portion of the roof, the homeowner files a claim. Because the claim is made while the policy is in effect, the insurance company pays to fix the damage, minus whatever deductible the homeowner's policy stipulates. Suppose, however, the homeowner fails to pay the annual premium. In this case, the insurance company will not pay to repair the damage because the policy was not in effect at the time of the claim. Even if the homeowner had paid the premiums for the last 20 years, it does not matter. Claims-made policies only provide coverage if the policy is in effect when the claim is made.

Claims-made insurance is common because it makes it easier for carriers to predict their costs. If professional liability insurance was occurrence-based, the cost of coverage would likely be prohibitive to most policyholders. The claims-basis approach allows the cost of coverage to be spread out over many years.

If a design firm decides to change insurance carriers, it still needs to maintain coverage for all its projects that were designed, constructed, and covered by the firm's previous insurance policy. To maintain this coverage, design firms need prior-acts liability insurance to cover the risks from earlier projects.

THE WELL-OFF AND THE NOT-SO-WELL-OFF

Design firms are not equal. Some are not as financially well-off and, consequently, have little to lose and much to gain. On the other hand, others are well-off and have much to lose. As we learned in Chapter 1, Daniel Bernoulli wrote that the well-off and the not-so-well-off approach risk differently. The latter are often more willing to take risks. This applies to design firms, as well.

When times are hard for design firms, there may not be enough work for everyone. Large firms have a somewhat easier time in smoothing out the roller coaster ride by having more employees, which allows them to take on more work and spread it out to more employees. If too much work comes in, it is easier to add a person to a staff of 50 than it is to a staff of four. In addition, larger firms can generally hold employees longer, if they want. Not all do, of course. Some large firms have a reputation as "hire-fire" companies. Different firms have different strategies for handling the risks associated with fluctuations in work loads. One method of managing the risk is to loan employees to other firms.

The AIA once had a standard agreement for the temporary loaning of employees. It was called AIA Document G813 titled *Standard Form of Agreement for the Temporary Placement of Employees.* The form was used for sharing employees between two design firms during slow periods. The form called the firm loaning the employee the *permanent employer* and the firm borrowing the employee was called the *temporary employer.* Based on the form, the temporary employer agreed to pay the permanent employer for the hours worked by the loaned employee at a rate large enough to cover the salary and salary-related expenses of the loaned employee. In addition, the temporary employer agreed not to "steal" the loaned employee for a period of six months after termination of the agreement.

Many firms used this form when loaning employees and a few still do. The agreement, however, is silent on which employer's professional liability insurance covers the work of the loaned employee. The agreement is clear that the temporary employer is not paying the permanent employer's full overhead costs, which include the cost of liability insurance. However, without any additional clauses in the agreement, it is likely that neither the permanent nor the temporary employer's insurance is providing professional liability coverage for the work of the temporary employee.

For smaller firms, employees on overhead for extended periods of time pose a great risk to the financial viability of the small firm. Letting key employees go and then finding new ones when the workload increases is also a risk and an added cost. For small firms, temporarily loaning employees in order to keep them busy, even if the liability insurance situation is fuzzy or outright compromised, is more important and outweighs the risk of what might happen if something goes wrong. For larger firms, the situation is generally reversed. Of more importance to larger firms is the clarity of insurance coverage. Ambiguity regarding insurance is a greater risk than the need to loan an employee or borrow a temporary one. Depending upon their size and well-being, different firms react to risks differently.

STANDARD OF CARE

According to USLegal.com's online legal dictionary, "Standard of care refers to the degree of attentiveness, caution, and prudence that a reasonable person in the same circumstance would exercise." For design professionals, this means that the standard of care is determined by what other design professionals would do under the same conditions. Owner/design professional agreements attempt to define the standard of care in clauses such as, "The design professional will endeavor to provide services in a manner consistent with the level of care and skill ordinarily exercised by other design professionals in the same locale." Another common definition is, "The design professional will strive to perform its services in accordance with a manner consistent with the level of care and skill ordinarily exercised by other design professionals in the same locale."

There are a few things worth noting: First, the meaning of standard of care is subjective and open to interpretation. Second, note the use of the word *endeavor* in the first agreement clause and *strive* in the second. The two words are roughly equivalent and suggest that the design professional will only *try* the best he or she can to perform in a manner consistent with the standard of care. A clause worth avoiding is, "The design professional warrants that it will provide services in accord with those provided by other design professionals in the same locale." Other equally poor choices would be to substitute the word *guarantee* or *insure* for *warrant.* Such words carry with them the duty to provide services equal to other design professionals, which increases risk because it is a promise the design professional may not be able to keep. Avoid agreement clauses that say that "services will be performed with the highest standard of care." There is a standard of care, and it is what it is. Do not take on an added risk by agreeing to a level of services that is subjectively greater than the already subjective standard level of care.

Note the use of the term *same locale* in both agreement clauses. What design professionals need to know in order to practice in a manner consistent with their peers varies between legal jurisdictions and geographic regions. For example, an architect might detail a shingled roof in Southern California differently than Alaska. An architect from Southern California who takes on a project in Alaska had better know how the typical local design professionals detail the roof because that will be the expected standard of care.

MANAGING LIABILITY EXPOSURE

Some projects and clients present higher liability risks than others. There are a number of factors to consider in reducing liability exposure. Some of the key factors are:

- Determine all of the design firm's insurance needs.
- Develop a strong and productive working relationship with an insurance broker or agent.
- Implement broker/agent risk management suggestions that are relevant and desirable.
- Avoid working with sue-happy clients.
- Use standard agreement forms whenever possible. (We will explore more about this later in the chapter.)
- Always be on the lookout for competent staff. Hire them, and constantly improve the quality of the staff.
- Train employees and encourage continuing education for employees.
- Implement a quality assurance/quality control (QA/QC) program.
- Avoid contract clauses that increase risk. (We will discuss this further later in the chapter.)

CLAIMS RESPONSE PLAN

Design professionals are judged by their actions. This means that they are judged by what they do, what they don't do, why they do it or not, how fast they do it, and whether other design professionals under similar circumstances would do the same thing. At some point, an owner may disagree with the actions of a design professional and strongly voice his or her dissatisfaction. Sometimes design professionals can defuse claims by resolving or paying to fix problems before they rise to the level of claims. Such preemptive actions can keep clients and save money. This can be particularly important in retaining good and repeat clients. Often a few thousand dollars wisely spent can save a client and head-off expensive, time-consuming, and nerve-racking claims.

It is the day no design professional looks forward to, but sooner or later, it may arrive: the day the design firm is hit with a claim. The first step is to notify the firm's insurance carrier. If the firm has an attorney, the attorney should be notified, as well. The attorney must be acceptable to the insurance carrier. If the design firm does not have an attorney, the insurance carrier will assign one.

The next step is to designate a firm spokesperson for the claim. All the relevant documents should be gathered because they may be subpoenaed. Generally, this includes all drawings, sketches, correspondence, specifications, and so on. It is good to discuss the claim with team participants and get them to write down their recollection of events. It is also essential to instruct the design team on appropriate conduct. In general, this means maintaining confidentiality, not talking to others about the claim, and if asked by others to comment, referring the inquirer to the appropriate team spokesperson. If applicable, it is a good idea to visit the site and take pictures. Above all, it is important to be honest, ethical, and to tell the truth.

I recall a case where an underground tank was installed in a hole that had not yet been backfilled. It buoyed up out of the hole after a big rain storm. As the project manager, I received a call from the owner's on-site representative who told me about the incident. He ended his phone call with me by saying that there was no need for me to come to the site. He and the contractor would work through the problem. Alarm bells went off in my head. When I hung up, I immediately went over to see our in-house structural engineer. The next day he was on a plane to the job site. I was concerned that without a representative of our firm present the contractor and owner might blame us simply because we had no one there to assess the situation and represent our point of view.

As it turned out, the rain filled the hole with water. On prior rainy occasions the tank had not lifted out of the hole because the contractor had kept the tank filled with water, thereby increasing its weight. For some reason, the day after this heavy storm, the contractor began pumping the water out of the tank. As the tank weight decreased, its buoyancy increased, which strained the cables and turnbuckles, eventually breaking one of them and, through a domino-effect, the others broke as well. The price of the plane ticket was well worth the expense because we were able to learn firsthand what had happened. After the structural engineer called me and explained the situation, I felt much better, knowing that we had not somehow been responsible for the mishap, and that we would be in a better position to defend ourselves because we had made our own assessment of the situation. An ounce of prevention was worth a pound of cure.

The following are general rules of conduct recommended by insurance carriers and attorneys. Their purpose is to preserve insurance coverage and to protect the firm's legal position.

- Maintain confidentiality.
- Follow all insurance policy requirements and procedures.
- Do not admit fault, even though it may seem tempting. This could void your insurance coverage.
- Do not agree to pay money or provide services for the claim, because this may also void insurance coverage.
- Do not publicly blame other parties, particularly the firm's consultants.

There is often a delay between the time the design professional's action or inaction took place and the time a claim is made. Sometimes many years go by. One of the reasons professional liability insurance policies are claims-made policies is because it is often difficult to know exactly when the act occurred that caused the resulting injury.

An event that precipitates a claim has three triggers. Generally all three must fire before an insurance company acknowledges that there is a claim against the policy. However, sometimes insurance companies will agree that a claim has been made against the design professional if only the first two triggers fire.

1. There must be an injury. Injury is a general term. It does not necessarily mean a broken leg or some other physical injury. The injury can also be financial. For example: "It costs too much money to keep the HVAC system running. It breaks all the time and it is expensive to fix and maintain." Alternatively, it could be a roof that leaks every rainy season, causing water damage to interior furnishings. These are financial injuries.
2. There must be an alleged wrongdoing. The claimant alleges that the design firm did something it shouldn't or didn't do something it should have. Because of this the injury occurred.
3. There must be a demand for compensation. Compensation can take the form of monetary compensation, meaning the design professional must pay some amount to the claimant. It could be a request for services, such as the services to fix the defect that led to the claim. It could also be a combination of the two—payment and services.

It is the design professional's responsibility to report the claim to the insurance carrier. Some insurance policies require notification upon "first knowledge." In this case, it is possible that only the first two triggers listed above have fired and the claimant has not yet decided about the third. Because design firms have to act relatively quickly, it is a good idea to have a plan in place for responding to claims. Principals of the design firm as well as project managers need to be aware of the response plan and its procedures. A claim response plan should address the following:

- Who should the project manager notify if a claim occurs?
- What are the claim-reporting procedures as stated in the insurance policy?
- Who within the firm is responsible for reporting the claim to the insurance company?
- What are the proper procedures for reporting a claim? It is all right to phone first and follow-up with a letter.
- Who is the firm's attorney? Most insurance policies require the design professional to use an attorney that is preapproved by the insurance carrier.
- Who within the design firm is responsible for claim management?

In addition to having a response plan in place, the following actions need to happen when there is a claim:

- Assemble all of the relevant documents.
- Notify former employees who worked on the project if their assistance is necessary or their testimony is required.
- Contact subconsultants who worked on the project. Consultants will have to defend themselves through their own insurance policy.
- Hire an in-house expert relevant for the particulars of the case. The claimant is certain to hire one. The design firm will need to rebut whatever the claimant's expert might say.

EVEN THE INNOCENT PAY

Some claims are "nuisance claims." It is clear that the design professional did nothing wrong, but a claim is filed nevertheless.

It must be defended, and that costs money. Insurance companies know that on occasion it is more expensive to fight a claim than it is to settle it, even when the design professional is innocent. Claims have a nuisance value, which is equal to the dollar difference between fighting the claim and winning compared to the cost of settling. Sometimes ultimately being right is more expensive than just making the claim go away with a compromise. In some instances, the design firm may want to fight because it knows it has done nothing wrong, but the insurance carrier wants to settle. Generally it is best to follow the insurance carrier's advice. If the design professional contests the claim and it ultimately costs more than settling, the design professional may have to pay the nuisance value of the claim.

PROFESSIONAL SERVICES CONTRACTS

Some of the risks that design professionals face are related to agreements between owners and design professionals. Design is challenging and risky without contractual risks. Fortunately, these can be controlled by following a few basic risk management rules:

1. Use standardized agreements whenever possible
2. Understand the agreement's provisions
3. Avoid language that increases risk
4. Use appropriate types of fees for services provided
5. Identify excluded as well as included services
6. Specify how disputes will be resolved

Use Standardized Agreements Whenever Possible

Risk is reduced by using tried and proven standardized agreements because they address and alleviate many known project uncertainties. The most widely used owner/design professional standardized agreements for the procurement of design services are the *Standard Form of Agreement Between Owner and Architect,* B141, published by the American Institute of Architects (AIA), and the *Standard Form of Agreement Between Owner and Engineer for Professional Services,* EJCDC 1910-1, published by the Engineers Joint Contract Documents Committee (EJCDC). The EJCDE is composed of members representing the American Society of Civil Engineers (ASCE), American Consulting Engineers Council (ACEC), and the National Society of Professional Engineers (NSPE).

The AIA's owner/architect agreement has been published for more than 100 years and has been updated and improved numerous times throughout its history. The EJCDC has published its owner/engineer agreement for more than 30 years with frequent updates. There are no other design services agreements with greater precedence or wider use. Both are written fairly, without partiality toward one party to the agreement or the other. They define the rights and responsibilities of both the owner and the design professional and have been tested many times in court.

Understand the Agreement's Provisions

Some design firms have developed their own standard agreements. Generally these are based on the AIA or EJCDC contracts, or on another professional organizations' agreements in combination with the firm's past experiences. These agreements tend to include clauses that specifically address risks that have threatened the design firm in the past. Because they do, the agreements are frequently weighted in favor of the design firm.

Many owners have their own agreement forms. These agreements are weighted, sometimes heavily, in the owner's favor. These agreements tend to treat design services like pieces of equipment or office supplies and the design professional like a contractor. Indeed, some refer to the design professional as the "contractor" within the agreement. Some even call their agreements "purchase orders." It is important that design professionals read and understand the provisions of the agreement. If they are unsure of the meaning of unfamiliar or conflicting clauses, they should seek legal advice.

Design firms should not start work without an agreement. Often this is tempting. Staff is available, the project manager is anxious to go, and, if the firm's work load is light, starting work on a project that's "a sure thing" sounds like a good idea all around. What can be the harm? The following example illustrates what can go wrong. It also shows what can happen if a design firm is unsure of the meaning of an agreement.

In the mid-1990s, business for a particular northeast paper company was booming. To keep up with demand, the company needed to expand its paper mill, so it hired a design firm. The firm sent an agreement to the paper company for signature. Among other provisions, the agreement said that the design firm would design the facility consistent with "good engineering practices," a rough equivalent to the design professional's normal standard of care. The paper company did not sign the agreement, however. Instead, it sent a purchase order to the design firm for it to sign. The purchase order was worded differently regarding the standard of care. It stated that "all materials and articles covered by this order…will be free from defects in material and/or workmanship, and merchantable." The design firm did not sign the purchase order, but went to work anyway. Eventually the paper company and the design firm signed a version of the design firm's original agreement. However, the agreement referenced the paper company's purchase order, which included the previous statement stipulating a higher standard of care. After the project was complete, the paper company filed a claim against the design firm stating that it was financially injured because the design firm did not perform to the standards of the agreement because some of the materials did not live up to the paper company's expectations. The case went to court and a jury concluded that the referenced higher standard of care clause in the paper company's purchase order, which the agreement referenced, superseded the clause in the design firm's agreement. Consequently, the jury found in favor of the paper company.

The lesson in the story is twofold. First, design professionals need to read and understand the provisions of the agreements, including the provisions of referenced documents in the agreement. Secondly, design firms should not start work until a clearly worded agreement is signed.[50]

Some owners claim that their agreements are not negotiable. They maintain that if the design firm wants the work then it must accept the agreement as is. Many years ago, a project manager for a major transportation company told me that if we wanted the work, we needed to sign the agreement without negotiation. It would be too much hassle and "take too long to get legal to agree to changes," he told me, so he didn't want to bother. I suggested that our company attorney talk directly to his company's attorney. He reluctantly agreed. A compromise was worked out, amenable to both attorneys. Generally, owners will negotiate because it is not in an owner's best interest to require levels of performance that transcend what is insurable by a design firm's liability insurance. Many owners know this upfront. Others can frequently be convinced of the wisdom, if the design firm or its attorney is willing to present the case fairly and offer a solution.

Avoid Language That Increases Risk

Some agreement provisions impart great risk. If the risk is too great, it can be a deal-breaker, and then it's best for the design firm to walk away from the project. Two common deal-breaker clauses that owners insert in agreements are:

- Indemnity clauses that require the design firm to indemnify (assure, underwrite, hold-harmless) and defend the owner against all claims, not just those due to the design firm's negligence
- Guarantees, warranties, or certification clauses that require the design firm to guarantee, warrant, or certify things that are beyond the firm's control

Some words add risk because they increase the responsibility of the design firm beyond the normal standard of care. The following words increase risk:

- insure
- ensure
- certify
- guarantee
- optimize
- maximize
- minimize
- supervise

As discussed in Chapter 4, design professionals should avoid using these words whenever discussing their capabilities. In addition, they should use caution when using them in agreements. They should seek legal advice before doing so.

Use Appropriate Fee Types for Services Provided

Some projects are better defined at the outset than others. Some are so vague that it is impossible to know exactly how long they will take to complete. Others are explained in considerable detail, with specific objectives, and program and schedule requirements. Because the definition of projects varies, it is important to choose a fee type that is appropriate for the degree of uncertainty in the project.

There are many different types of fee structures used in the contracting of design services. The five most common types of fees are:

1. Lump sum (LS), also called fixed-price
2. Time and materials (T&M)
3. T&M with a maximum not-to-exceed
4. Cost plus fixed fee
5. Cost plus fixed fee with a guaranteed maximum price

Lump Sum. Lump-sum agreements are fixed-price agreements. The design firm agrees to perform all of the work for an agreed-to lump-sum price. With this type of contract, any money not spent by the design firm is profit. If the firm spends more than the lump-sum amount, the project loses money. Payment is based on contract-defined deliverables, often on the percentage basis—if the project is 35 percent complete, the firm invoices for 35 percent of the lump-sum amount. Lump-sum contracts are best used when the design project is well defined. If not, lump sums can be risky because poorly defined projects tend to take more time and cost more to complete than can be anticipated at the outset.

Time and Materials. With T&M agreements, the design firm is paid for time and materials spent. There is no maximum price with T&M agreements. Whatever the design firm spends, the owner pays. Generally, hourly labor rates are negotiated and agreed to upfront. The rates are comprised of staff salary rates, salar-related expenses, and profit. They vary depending upon the salary and experience levels of different staff members. Some T&M agreements allow design firms to add a service charge to consultants equal to a negotiated percentage—commonly a 5 percent and 10 percent markup. Generally costs such as printing and postage are not marked up.

T&M agreements tend to be very good agreements for design firms because it is nearly impossible to go over budget. There is no budget; the client pays for every labor hour spent. As long as the negotiated rates are enough to cover a design firm's costs and include a reasonable profit, T&M agreements have very little risk. However, unlike lump-sum agreements, it is impossible to improve a project's profit margin with T&M agreements because the percent of profit is fixed into the hourly rates.

T&M agreements are generally used when it is difficult or impossible to determine a scope of work in advance. They are a bit more difficult to administer than lump-sum agreements because invoices require backup that show labor-hours spent and direct costs incurred. If some direct costs or labor hours are challenged by the client, payment can be held up. Sometimes challenged costs are denied outright and payment for them is never received, which has an adverse affect on a project's profitability. With T&M agreements, there is no way to make up for such losses.

Time and Materials with a Maximum-Not-to-Exceed. To owners, straight T&M agreements are like blank checks. Some feel that there is little incentive for the design firm to be efficient with T&M agreements. Time and materials with a maximum-not-to-exceed solves this problem. With this agreement, the firm is paid for labor hours and other direct costs (ODCs) until it reaches the maximum dollar amount. Unless there is a change in the scope of work, the agreement has a ceiling, a dollar cap that cannot be exceeded. Many owners also prefer this contract to lump-sum agreements because it eliminates the possibility of the design firm receiving a windfall profit should it take far less money to finish the project than originally agreed upon. These agreements pose little risk to an owner and are consequently very popular. For the design firm, some pretty exacting financial management is required on the part of project managers. To maximize profit, the project must spend as much of the fee as possible, without going over, which is nearly impossible. Exceeding the maximum dollar amount cuts into profit, and running under adversely affects profitability, too, because the profit associated with the hours not spent is not realized.

Cost Plus Fixed Fee. With a cost plus fixed fee agreement, the design firm is paid for all its costs, labor, and materials. However, the labor costs do not include profit. Profit is a negotiated fixed amount that does not vary with actual costs. With this type of contract, the amount of labor and ODCs required to accomplish the specified work is estimated. The actual cost may vary, but the profit is fixed. If it takes more hours or costs to accomplish the work, the firm is paid for the labor and ODCs, but the profit does not change.

This fee agreement has advantages for both owner and design firm. The design firm has every incentive to run the project effectively because the percent of profit increases if fewer hours are spent. Should the project go over budget, the design firm loses profit as a percentage of total costs. While the owner is out additional costs, at least he or she is not paying more in profit.

Cost Plus Fixed Fee with a Guaranteed Maximum Price. This type of fee agreement is a variation of the cost plus fixed fee contract previously described, except that in this case, the agreement has a guaranteed maximum price. If the design firm goes over the maximum amount, it is not compensated for its costs. This contract type has all the advantages of the cost plus fixed fee contract, but protects the owner from severe cost overruns.

Identify Excluded As Well As Included Services

Agreements include descriptions of the scope of services required. Equally important, but often overlooked, are descriptions of work that is excluded. Identifying excluded work reduces risk because it lessens the chance of misunderstandings and false expectations. Owners may assume certain services are included because they were included previously or because owners mistakenly assume they were normally provided design services. For example, some projects include programming services, which are generally not basic design services but rather additional services, although commonly provided. Some projects require surveys, others do not. In some cases the owner provides the survey, in others, the design professional does. The same is true for geotechnical investigations, traffic studies, and a host of other services design professionals can provide but do not unless specifically asked to do so. If there is ambiguity regarding certain services, the agreement should make clear which services the design firm will provide and which services it will not.

Specify How Disputes Will Be Resolved

Lawsuits can be very stressful. They expose the design firm to enormous financial uncertainty, which dramatically increases risk. Lawsuits can be stressful to clients, as well, for the very same reason. Consequently, design professionals, owners, and contractors sometimes employ alternative dispute resolution techniques. Three alternative dispute resolution techniques worth considering are:

- Partnering
- Arbitration
- Mediation

Partnering. Partnering is a series of formal team-building sessions, usually led by an independent facilitator. It is used to encourage a team approach between owners and design professionals to problem solving, rather than an adversarial approach. It is frequently used during construction projects to encourage cooperation among team members representing the owner, design firm, and the contractor, who often have disparate and competing interests. The goals of partnering are:

- Demonstrating that working together harmoniously is more likely to bring success than a confrontational approach
- Emphasizing the common goals of the team members and encourages an appreciation for the divergent experiences and expertise that each team member brings to the project
- Improving communication, promoting fair lay, and eliminating the "us-against-them" mentality
- Preventing disputes that can lead to hostility, possible financial injury, and eventually to lawsuits

The first session generally takes place before the project begins. Subsequent sessions take place at regular intervals throughout the life of the project. The sessions include various exercises that promote cooperative problem solving and team building, in effect a collaborative spirit. Collaboration permits a team to leverage its skills, time, and resources for the overall benefit of the group and project. Successful partnering sessions accomplish the following:

- Commitment from team members to cooperate
- Establish effective, open, and honest communication
- Create mutually acceptable procedures for resolution of project-related issues and problems
- Summarize the team's objectives for the project in a partnering mission statement
- Cultivate a win-win problem-solving attitude
- Develop metrics for measuring whether the project objectives are being met
- Encourage creative ways to control costs
- Establish a follow-up process for fine-tuning procedures

Partnering is successful only when all team members embrace the process and actively try to make it work. If partnering is treated as just another exercise, its effectiveness is greatly diminished.

Arbitration. The most formal of the three dispute methods is arbitration. In arbitration, opposing parties and their attorneys present their cases to one or more individuals who are empowered to render a decision. Most arbitrators for design- and construction-related disputes are architects, engineers, and constructors, so they are usually familiar with the technical issues that frequently bear on the resolution of these disputes. Most arbitrators receive training from and are members of the American Arbitration Association (AAA), an organization with a long history of resolving disputes.

Agreement clauses that stipulate arbitration for dispute resolution often call for binding arbitration. Decisions rendered through binding arbitration are enforceable by the courts. Arbitration does not follow the same rules of evidence as civil courts. Arbitrators can render judgments that are not based on legal principles. They can and do permit hearsay testimony and often allow the admission of irrelevant testimony and evidence. They do not have to abide by the terms of the design or construction agreement, and they are not required to state the reasons behind their decisions.

The AIA used to recommend arbitration as the preferred dispute remedy. However, in light of the shortcomings of arbitration as just described, the AIA changed its position in the mid-1990s. It modified its *Standard Form of Agreement Between Owner and Architect* to recommend mediation as the preferred form of dispute resolution.

Mediation. Mediation is a voluntary method of resolving disputes. Parties agree in advance to resolve disputes themselves. To assist the parties, an independent mediator acts as a go-between, helping the parties find an acceptable resolution. Mediators work purely as neutral facilitators. They do not function as attorneys, experts, or advisors to the parties. The mediator's role is to assist the parties in developing a comprehensive understanding of the dispute by taking into account each other's perspective. From there the mediator helps the parties work toward an amenable, practical, and lasting resolution.

Mediation is usually not binding. The parties do not have to accept the resolution and the resolution is not enforceable by a court of law. The parties can always file a lawsuit as a last resort. However, parties generally accept the outcome. Sometimes the agreed-to resolution is simply documented, such as in a memorandum of understanding. Other times, it may result in modifications to an existing agreement or a new legally binding agreement.

WORKING FOR FREE

Recall the architect Charles Bulfinch and how his tontine partnership situation forced him to work for free. Today the same can happen to design professionals if they do not pay attention to agreement clauses. A common clause is worded something like this:

> The Owner reserves the right to reject all or any portion of an invoice. If the Owner disputes the Design Professional's invoice for any reason, the Owner may withhold the disputed amount until it is resolved and the Design Professional shall continue to perform services as set forth in this Contract without interruption.[51]

This clause may have been well intentioned when it was first written. Design professionals do get a bit lax from time to time in attaching the necessary backup to invoices so that owners can understand all of the charges. However, what this clause says is that, regardless of the reason for the dispute, the owner does not have to pay until the dispute is resolved. Nevertheless, the design professional must keep working.

A variation of this clause goes something like this:

> In the event of a dispute between the Owner and the Design Professional, the Owner and Design Professional agree to continue performing their respective obliga-

> tions without interruption or delay until the dispute is resolved.[52]

On the surface, this clause sounds innocuous. Yet suppose the dispute is over the amount or entitlement of payment to the design professional and the owner decides not to pay? In this case, the clause stipulates that the owner only needs to perform all his or her other agreement duties. The owner does not have to pay the design professional because the payment is the item of dispute. The design professional, however, must continue to perform all of his or her obligations.

A more onerous clause is:

> The Design Professional acknowledges that a stoppage of the work could cause the Owner actual or consequential damages. Any dispute that arises between the Owner and the Design Professional shall not become the basis for the Design Professional to delay or stop work. Should the Design Professional delay or stop work, the Owner shall be entitled to recover all actual and consequential damages from the Design Professional that result from the Design Professional's delay or stoppage of the work.[53]

These provisions or variations of them have been tested in various state courts and upheld. Unscrupulous owners can use them to avoid paying design professionals for services rendered and force them to continue work. Disputes for small sums of money might not be terribly disruptive to large design firms, but for smaller firms the results can be very damaging. It is important to recognize potential risks by carefully reading agreements. When in doubt over the meaning of an agreement clause the firm's attorney should read the agreement.

Remember that all agreements are theoretically negotiable. I say theoretically because some owners claim that theirs are not. They rationalize that other firms haven't had a problem with their agreements or its clauses, so why should you? In these situations, the design firm must make a business and risk management decision. Is the firm willing to accept the clause and its consequences? Is this the kind of client the firm really wants to work for?

CHAPTER SUMMARY

Design professionals do not produce a product—they provide a service. Because of this distinction, product liability does not apply to their services. Instead, professional liability applies.

Professional liability requires design professionals to practice in a manner similar to the way average, competent design professionals would practice under the same circumstances, within the same legal jurisdiction. This establishes what is called the "standard of care." Design professionals do not have to be perfect; they only need to meet the requirements of the accepted standard of care.

Design professionals buy professional liability insurance in order to manage risk and protect themselves from so-called negligent acts—also called errors and omissions—that may arise from providing their services. Liability insurance manages risk by transferring financial risk, although not all of it, to the insurance carrier.

Liability insurance is sold by insurance company agents and by independent insurance brokers. Insurance agents work directly for insurance companies and only sell insurance policies provided by the particular company they represent. Independent brokers sell insurance policies from a variety of companies, but may have limited access to certain insurance companies and policies.

Insurance policies should not be selected based on premium costs alone. Factors to consider are:

- Cost relative to the coverage of the insurance policy
- Experience of the insurance company writing professional liability policies

- Ability of the insurance carrier to provide coverage in the geographical areas where the design firm does work
- Financial strength of the carrier
- Whether the insurance company is an admitted carrier
- Willingness of the carrier to write additional liability policies
- Comprehensiveness of the policy—does it cover all of the firm's principals and employees

Design firms should develop a claim response plan, which should:

- Identify who is responsible for notifying the insurance carrier
- Identify who within the firm will be the claims spokesperson and coordinator
- Adhere to all insurance policy requirements
- Stress the importance of confidentiality, honesty, and professional integrity in dealing with the claim

Risk is reduced when the fee type is appropriate for services provided and the project definition. There are five basic types of fees common to design:

- Lump sum, also called fixed-price
- Time and materials
- T&M with a maximum not-to-exceed
- Cost plus fixed fee
- Cost plus fixed fee with a guaranteed maximum price

The method of dispute resolution should be specified in the agreement. Three alternative dispute resolution techniques are worth considering. They are:

- Partnering
- Arbitration
- Mediation

Chapter

8

RISK AND CONSTRUCTION SUPPORT SERVICES

The Hyatt Regency Hotel in Kansas City, Missouri, opened in July 1980. The 750-room hotel's main architectural feature was its splendid atrium lobby. The 50-foot-tall atrium served as a connector between the hotel's guest tower on the north side of the lobby and the dining and conference facilities on the south side. Suspended on 1-1/4 inch rods from the atrium's roof trusses, three 120-foot-long elevated steel walkways passed through the lobby connecting the guest tower with the conference facilities. One walkway was suspended along the east side of the lobby, connecting the third floor of the guest tower with the conference facilities. On the west side, above the hotel's cocktail lounge, a double-decker steel walkway connected the second and fourth floors.

During construction, the contractor submitted shop drawings of the suspended walkways to the architect for review. The architect forwarded them to the structural engineer. The shop drawings for the double-decker walkway included a subtle but very important difference from the contract drawings. The disparity was the connection of the suspension rods and the transverse box-shaped beams that supported the upper walkway. At each suspension rod and box beam connection, the contract drawings showed a single rod passing through the box beam. The contractor's shop drawings showed two rods. One rod ran from each box beam to the roof structure above and another ran between the upper walkway and the lower walkway. In the contract drawings, the box beams of the upper walkway supported *only the load of the upper walkway.* In the contractor's shop drawings, the box beams of the upper walkway *supported the load of both walkways—twice the load.* The shop drawing was reviewed by the structural engineer, returned to the architect, and returned to the contractor stamped "approved."

On Friday night, July 17, 1981, one year after the hotel first opened, over 1,600 guests were in the atrium lobby, dancing and listening to a live band. By chance, the event was being videotaped. The video showed 63 people standing on the double-decker walkway, most of them located near the south end of the walkway, where the view of the band was better. Together, the 63 people created a loading condition that was only a fraction of the design load required by the building code.

Suddenly, a loud bang rang out. The double-decker walkway fell, shattering windows and breaking fire sprinkler piping as it crashed to the lobby floor. One hundred and fourteen people were killed and over 200 injured in what was the worst structural failure in the United States.[(54)]

In the subsequent investigation by the National Bureau of Standards, it was determined that the structural failure first occurred at one of the upper walkway suspension rod and box beam connections. When it failed, other connections quickly became overstressed, and the walkway failed in a chain reaction, or what is called a progressive failure. The investigation brought the contractor's shop drawing change to light. The investigation revealed that the structural engineer reviewing the shop drawings failed to do the necessary calculations to verify that the contractor's change could handle the code-mandated design loads. In addition, the investigation determined that the original design was also inadequate; it did not meet the loading requirements of the building code. Although inadequate by code standards, the original design was stronger than the contractor's design, suggesting that under the loading conditions that led to the failure of the walkway, the original design might not have collapsed. In the aftermath of the investigation, the structural engineering firm's principal and project manager both lost their Missouri state engineering licenses.

SERVICES DURING CONSTRUCTION

As the previous story illustrates, the review of shop drawings and other contractor submittals is a serious responsibility. In addition to submittal reviews, there are other standard services that design professionals provide during construction that are equally important. Alternatively called contract administration, construction administration, or construction support services, these services carry with them various forms and degrees of risk. Collectively, the services assist the owner in administering the construction contract with the contractor. Construction support services frequently include:

- Submittal reviews
- Construction procurement services
- Construction observation
- Responding to RFIs (requests for information)
- Writing change orders
- Review contractors' requests for pay
- Project closeout

Submittal Reviews

As the Hyatt Regency disaster illustrated, there can be grave consequences if submittals are not reviewed correctly. Sometimes these problems do not surface until much later, even years after the facility is in use. If submittals are not accurately reviewed, materials may not fit together properly, systems might not work, and joints or details might leak. In extreme cases, structural components might fail causing injury or death.

Contractors' submittals take the form of shop drawings, product literature, physical samples, color swatches, reports from testing agencies, warranties, and operating instructions and maintenance manuals. All must be thoroughly reviewed for conformance with the contract drawings and specifications. Submittal review is the last chance for the design professional to see, review, and think about how the various parts that form the complete facility go together and interrelate. Contractors make mistakes, too. They miss things, or misinterpret the contract documents. If there are any shortcomings or problems, this is the last chance to catch them before paper problems become physical problems.

Because the review of submittals is so important, design firms should establish standard procedures for reviewing them. If the submittal review process is left to the discretion and idiosyncrasies of different project managers, the process and its effectiveness as a risk management tool will vary considerably from project to project. Invariably, some PMs will take the process less seriously than others, which increases risk for the design firm.

The following list of procedures has proven effective in reducing risk during submittal reviews. Collectively, they describe the major components of a comprehensive submittal review plan.

- Every project manager is responsible for following the design firm's standardized submittal review procedures.
- Every project manager selects a person within the firm to be the submittal-review coordinator. On small projects this might be the project manager.
- All project submittals come into the office through the submittal-review coordinator, are tracked by the coordinator, and leave through the coordinator.
- Submittals are stamped with the date they are received.

- The coordinator keeps a log of all submittals, commonly called a shop drawing log. At a minimum the log includes the date the submittal was received, date it was distributed to the in-house reviewer or sent on to the consultant, date it was received back from the reviewer or consultant, review action taken, and date it was returned to the contractor.
- A standardized shop drawing stamp should be created. The stamp contains check boxes for the actions taken, for example, "approved," "make corrections noted," "amend and resubmit," and so on. The names of the possible actions vary depending upon the design firm's insurance carrier. If in doubt, consult with the firm's insurance carrier.
- A standardized submittal review transmittal letter should be developed. The letter includes check boxes for the actions taken along with the following, at a minimum: space for review comments, submittal number, specification section reference number, name of submittal, name of contractor, subcontractor, or material supplier.
- Consider adding a standardized explanation clause to either the shop drawing stamp or the submittal review transmittal letter. The clause explains both the purpose and limitations of the review and thereby reduces risk (see the example following this list).
- Establish filing procedures for submittals. After the project is complete, the submittals should be retained for an established period. Consult with an attorney or insurance carrier for the recommended length of time.
- Establish submittal conformance procedures. Generally contractors submit multiple copies of a submittal. The design professional reviews the submittal, marks up one copy, and the comments are then transferred to other copies by a staff member designated as the "conformer." Conformers must be conscientious with excellent work habits so that review comments are transferred accurately to all copies.

- All submittals should be checked using the firm's quality control program. In addition to checking for accuracy, the quality control reviewer checks to see that required documentation that supports the review is present.

The following clause is a slightly modified and shortened version of a suggested shop drawing stamp statement contained in the Design Professionals Insurance Companies' *DPIC's Contract Guide.* The clause provides a good summary of the purpose and limitations of a design professional's submittal review.

> This submittal review is for general conformance with the design concept and information contained in the Contract Documents. Corrections or comments made on the submittal during this review do not relieve the contractor from compliance with the requirements of the Contract Documents. Approval of a specific item does not mean approval of the assembly of which the item is a com-

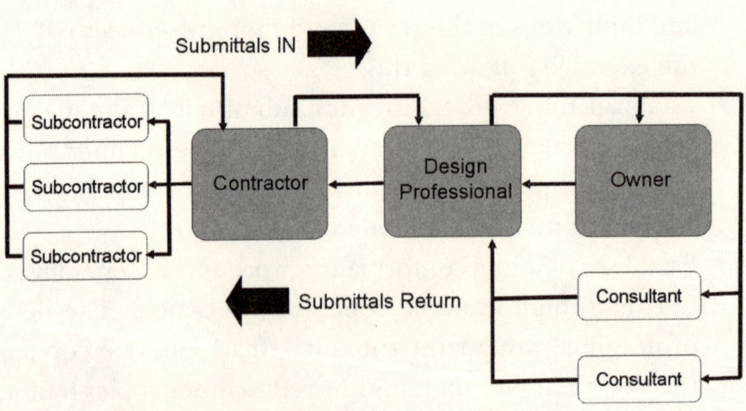

FIGURE 8.1 *Submittal review process.*

ponent. The contractor is responsible for dimensions, information required for fabrication, means, methods, techniques, sequences, procedures of construction, and performing all work in a safe and satisfactory manner.[55]

Figures 8.1 and 8.2 illustrate the submittal-review process.

With procedures in place to receive, record, distribute, track, review for quality, file, and return submittals, the design firm can quickly determine the whereabouts and actions taken on all submittals for projects currently under construction. Submittal-review guidelines should be established, as well. Following is a list of guidelines to follow:

- Prior to construction, provide the contractor with a list of required submittals.
- Require contractors to review, coordinate, and approve subcontractor submittals before they are submitted to the design professional for review.

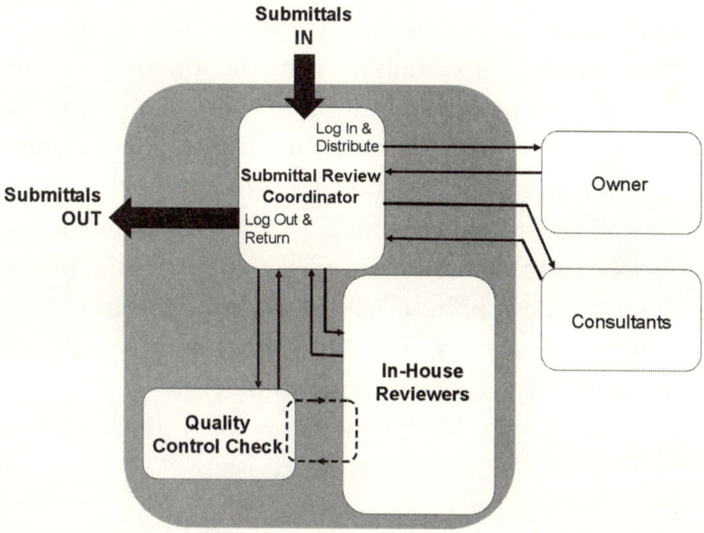

FIGURE 8.2 *Detail of the submittal-review process within the design firm.*

- Require contractors to bring substitutions and deviations to the contract to the design professional's attention in the submittal transmittal letter.
- Do not review submittals if crucial information is missing. If it is missing, return the submittal to the contractor and request resubmission with all pertinent information. Imagine the difference had the Hyatt Regency project submittal coordinator returned the shop drawings to the contractor with a request for calculations in support of the contractor's design change before reviewing the submittal.
- Review submittals in a timely manner. Construction contracts usually state how many days the design professional has to review submittals. Know the time period, and return submittals within the time frame. Many contractors have won claims for delay because design professionals have taken too long to review submittals.
- Reject shop drawings for substantive reasons only. An example follows this list.
- Do not accept substitutions that are not equal to the product or material specified. More about this follows.
- Do not review submittals for construction methods, means, or sequencing, as these are the sole responsibility of the contractor. Remember that the purpose of shop drawing review is to review submittals for design-related issues and compliance to the contract documents, not for the various procedures the contractor uses to construct the project.
- Do not review submittals for temporary aspects of the construction process, such as shoring or scaffolding, as these are the contractor's sole responsibility.
- Do not change the scope of work during a submittal review.

Occasionally the design professional unintentionally changes the contractor's scope of work during the submittal review process. Other times the change is done deliberately when the design professional sees that something is incorrect or missing from the design and fixes it on the shop drawing, thus changing the contractor's scope of work. Instead of making the change during the submittal process, a change order request (request for quotation) should be issued to the contractor.

As stated, submittals should only be rejected for substantive reasons. I remember a submittal for a state-of-the-art scientific computer research facility. The submittal was highly technical and included a series of extremely accurate electronic instrumentation measurements. The submittal was forwarded to the owner for review. A NASA rocket scientist reviewed the submittal. The owner rejected the submittal and returned it to the design firm and contractor. The contractor asked why the submittal was rejected. After a few days, the owner's reviewer replied that she rejected it because the average of the instrumentation readings had as many significant figures as the individual readings. She explained that when a series of numbers with three significant figures right of the decimal point are averaged, the resulting average can only have two significant figures. Because the contractor's average had three, it was rejected. Upon hearing her explanation, the face of the owner's project manager turned purple. The contractor just laughed. Moral of the story: reject submittals for substantive reasons, not for pedantic or petty ones.

When projects are competitively bid, contractors, subcontractors, and material suppliers read the specifications and base their bids on the assumption that the type and quality of materials and equipment specified is what is expected. The entire process of bid, contractor selection, submittal review, and construction is based on the honor system. The subcontractors and contractors base their bids on their honest and best judgment of what materials, equipment, and quality of construction is expected. When the bids are opened, the assumption is that the apparent low bidder bid on the same work and quality of work as all the other contractors. When the submittals come in, it is the design professional's responsibility to enforce the contract documents.

Substitutions that are unequal to the materials and equipment specified should be rejected. If they are not, it not only lessens the quality of the project, it also undermines the whole process. The bidding process becomes a game to determine which contractor is willing to take the greater risk that the design professional will accept materials and equipment of lesser quality.

Construction Procurement Services

Owners often ask design professionals how best to select a contractor and build a project. There are a number of construction procurement alternatives, also called project delivery systems. Alternative delivery systems have different advantages and disadvantages and consequently affect risk in different ways. The more common project delivery systems are:

- Negotiated sole source
- Bid by prequalified contractors
- Competitively bid
- Construction management
- Construction management at risk
- Design/build

Negotiated Sole Source. This delivery system involves asking only one qualified contractor for a proposal or price. The advantage is that the qualifications of the contractor are generally well known. The contractor may have built something for the owner before and established a good working relationship. Alternatively, the contractor may have come highly recommended by other owners, or the design professional may have had good experience with the contractor. That the contractor is a known commodity reduces risk. However, receiving only one bid generally increases the construction cost because the contractor has little incentive to keep costs down. If the contractor receives a high bid from a particular subcontractor, he or she may not feel the necessity to get additional bids from competing subcontractors. This increases the probability of the project exceeding its budget.

Bid by Prequalified Contractors. In this system, a select list of prequalified contractors bid the project. Design professionals are often asked to assist owners in identifying qualified contractors. Sometimes interviews are conducted, other times contractors are requested to submit their qualifications, and on still other occasions both are required. A list of three to six—rarely more—qualified bidders is established and only these bidders are asked to submit bids for the project. Because only prequalified contractors are permitted to bid, the owner has some control over which contractor will be selected. Customarily, the project is awarded to the lowest bidder, because presumably all are equally qualified. Owners who do not select the lowest bidder may find their selection process criticized and challenged by the qualified contractor who was the low bidder. To avoid this, many sophisticated owners always select the lowest bidder to keep the process transparent and fair to all participants.

Competitively Bid. In this process, the project is publicly advertised. Most public agencies must use this project delivery system by law, and many publicly traded corporations prefer to use it as well. All contractors are eligible to compete, provided they meet the licensing laws of the jurisdiction and specifically stated criteria contained in the project's bidding documents. Bids are received at a specific time and date, opened, and read publicly. The lowest bidder wins, assuming that all required supporting paperwork is in order. This delivery system tends to yield the lowest bid price. In exchange, the owner has almost no control over who will construct the project. With public bidding, the qualifications and sophistication of the bidders varies considerably, which increase risk for all parties—owner, design professional, and contractor. If, for some reason, an owner chooses not to select the lowest bidder, it is nearly certain that the lowest bidder will protest the selection. If this happens, start of construction can be delayed for a considerable time until the protest is resolved.

Another risk to competitive bidding occurs when a contractor inadvertently leaves some costs out of its bid. I remember a project where the low bidder was lower than all the others by a considerable sum. At the time, the economy was slow and contractors were hungry. Nine contractors bid on the project. The low bidder was nearly 25 percent lower than the next closest bidder. The next closest bidder and the seven other bidders were relatively close to each other and within the range of the design professional's estimate of probable construction cost. While the low bidder was happy to be low, he was concerned that he was too low. Perhaps there was a "bust" in his bid because he made a mistake and left something out. The owner and design professional were also worried.

One way of determining whether a low bid is too low is to determine its degree of deviation from the rest of the bids, and see if it is statistically significant. Recall from Chapter 1 the discussion regarding standard deviations. Remember that 95 percent of all samples fall within two standard deviations of the mean. If the lowest bid is farther away than two standard deviations from the mean, it is an outlier and this strongly suggests that something might be wrong. Study the bid in detail and compare it to the other bids. Ask if bids from all the major trades are accounted for? Is there an arithmetic error?

Of course, it could be that the contractor is low because he figured out a better and more economical way of building the project. That is unlikely if the bid's deviation from the mean is too great. It is much more likely that something was left out of the bid or that the bid contains some other error. In the situation previously described, the contractor reviewed his bid and withdrew, claiming an arithmetic error.

Bids that are too low not only create risk for the contractor, they also create risk for the owner and design professional. Contractors who leave too much money on the table frequently find ways to supplement their meager construction budgets with change orders. They look for things wrong or ambiguous with the construction documents. This increases the owner's risk, because it can lead to too many change orders, which could cause the project to exceed the owner's budget. For the design professional, risk increases, too, because at some point the number of change orders may cause the owner to turn against the design professional.

Construction Management. With this project delivery system, a construction management (CM) firm is selected to deliver the project, not a general contractor. The CM firm can be selected using any of the methods used to select design services. The CM bids various parts of the job in different bid packages. For example, the concrete foundation may be a bid package, the structural steel another, and architectural finishes another. The number of packages varies depending upon the project, the CM firm, and other factors. In construction management, the CM does not perform any work directly. Instead, it manages the multiple construction contracts and provides oversight. For example, instead of a contractor's superintendent, the CM firm has an on-site contracts manager who manages the daily activities and arranges to have the various contractors arrive and perform their work on schedule and in an organized and coordinated manner. The construction manager can competitively bid the portions of work or sole source work. The CM is not responsible for the overall project cost. The owner is. The various contractors all have separate contracts directly with the owner, not the CM. With construction management, the owner does not necessarily know how much a project will cost until all the contracts are let. Because contracts are frequently let throughout the project, it is difficult to know how much the project will end up costing until well into construction. For some owners, this is an acceptable risk, but for many owners this is not acceptable.

Construction Management at Risk. A variation of the construction management contract is CM at risk. In this system, the construction management firm must have a contractor's license because the CM holds the construction contracts. In CM at risk, the CM firm gives the owner a guaranteed maximum price for delivery of the project. The CM firm bids all the components as previously discussed and again, the CM performs no work itself, only supervision and oversight management personnel. For the owner, the project has a fixed price, which reduces the owner's risk. In addition, the owner can select the CM firm using a combination of qualifications and price.

Design/Build. This process takes CM at risk one step further. In this system, the design/build firm gives the owner a guaranteed maximum price for both the design and construction of the project. Many design/build entities are really multiple firms. Frequently they are a general contractor working in association with a design firm(s) as a consultant. On other occasions they are joint ventures. Some large design firms are also constructors with a contracting license. Some design/build firms self-perform some work, meaning they use their own employees to perform construction work. Others do not. With design/build, owners know before the project is even designed how much it will cost, because both the design and construction costs are determined early in the development of the project. This has great appeal for many owners.

Construction Observation

Design professionals do not direct construction work. They are not responsible for the methods and means of construction, nor are they responsible for construction safety. Instead, the design professional's role during construction is to observe construction in order to satisfy the owner that the work is being performed in accordance with the construction drawings and specifications. The contractor is responsible for construction means, methods, and safety. This is true so long as there is nothing in the agreement between the owner and the design professional that changes this arrangement or makes it ambiguous.

Normally design professional construction administration or construction support agreements have a paragraph similar to the following clause, which comes from the AIA's **Document B141** *Standard Form of Architect's Services: Design and Contract Administration.*

> The Architect, as a representative of the Owner, shall visit the site at intervals...to become generally familiar with and to keep the Owner informed about the progress and quality of the...Work completed...to guard the Owner against defects and deficiencies in the Work, and...to determine in general if the Work is being performed in a manner indicating that the Work...will be in accordance with the Contract Documents.... The Architect shall neither have control over or charge of...construction means, methods...or for safety.[56]

During construction, design professionals are frequently asked to provide construction observation services. This often involves regular trips to the job site to see how work is progressing. The words used to explain these services carry very different levels of expectation and risk. Phrases like "observation of the work" are much preferred to others such as "inspect of the work." The former suggests that the design professional is visiting the site to become generally familiar with the progress and/or to keep the owner advised. The term *inspection* carries with it the meaning that the design professional is doing detailed checking of the work and making sure that it is installed correctly.

The level of risk associated with these two services is dramatically different. Avoid clauses like, "the design professional will inspect the work" or "the design professional will ensure that the work is being installed in conformance with the contract documents," or "the design professional will certify that the work was installed according to the drawings and specifications." If the design professional inspects, ensures, or certifies the work, he or she crosses the line into the realm of the contractor and may be agreeing to provide services that are not insurable under the design professional's liability insurance policy.

Owners occasionally request that the design professional provide construction field personnel or inspectors. Providing field inspectors has advantages and risks. The advantages are:

- Field personnel work for the design firm and consequently have the same objectives as the design firm.
- Communication between the design office and field personnel is good, open, and honest.
- Field personnel serve as a second set of eyes for the design firm, able to catch mistakes before they happen.
- Representation of the firm to the owner is consistent.

There can also be disadvantages that can increase risk. They are:

- The likelihood of short-circuiting the submittal review process is increased.
- Submittal documentation can become lax.
- The contractor may perceive that the design firm's representative is biased toward the design team.
- The owner may perceive that the design firm field representative is biased toward the design office.

Many years ago, I worked on a project where the contractor arrived with truck loads of gravel to be placed beneath a concrete slab-on-grade. The slab was susceptible to moisture because of the high water table. The contract documents called for free-draining gravel, 3/4-inch rock to serve as a capillary break. No submittal was made. When the trucks arrived at the job site, the contractor asked the field inspector to review the sieve analysis for the gravel. To an experienced submittal reviewer, it would have been clear from the sieve analysis that what the contractor was proposing to use was not free-draining gravel. Instead it was material commonly used as aggregate base for road beds. With the contractor in a hurry and dump truck after dump truck filled with the stuff waiting in the street, the inspector approved the material.

Once the approved submittal made its way to the office, the experienced reviewer saw the mistake. It cost thousands of dollars to remove the material, a cost that was eventually paid by the design firm.

Responding to RFIs

Throughout construction, contractors and subcontractors have questions regarding the contract documents. Years of experience have shown that it is better for all parties concerned if the questions and answers are in writing. The written questions and answers provide a record trail of the information transfer. In general, the following procedures reduce risk during this important transaction.

At the beginning of construction, set up a standard procedure for RFIs. Often this involves the preparation of a standard RFI form and determination of the proper channels for submitting them. Contractors may have questions themselves or the questions may originate from their subcontractors. Often subcontractors write their RFIs in the field and hand them to the contractor's superintendent for transfer up the chain. The contractor's construction clerk copies the question onto the agreed-upon RFI form. Often the question is transferred verbatim; other times it is rewritten to make it clearer. The RFI is then sent to the design professional for a response. RFIs may arrive via mail, fax, or email. As one might expect, email is becoming a more and more popular form of communication. If the RFI includes a sketch or a portion of a drawing, the email may include attachments, such as PDF files.

Like submittals, RFIs should be logged in and out by the design firm's project coordinator. The RFIs are routed to the appropriate design professional for review by the project coordinator. Occasionally, the RFI involves more than one discipline, in which case the RFI may be routed concurrently or sequentially to the appropriate disciplines. Regardless, the answer to the RFI must be coordinated and clearly convey the appropriate answer.

The design professional provides a written response to the RFI. Sometimes this includes a drawing or sketch to communicate the answer better. The RFI is returned to the project coordinator, who reads and forwards it to the project manager for his or her review. The RFI is then forwarded to the project's designated quality control reviewer. When the reviewer finishes his or her review and concerns are adequately addressed, the project coordinator logs the RFI out and it is sent back to the contractor.

The following are some good risk management rules to follow when reviewing RFIs:

- All RFIs must be in writing.
- All RFIs must come from the general contractor. Subcontractors must channel their RFIs through the contractor.
- All RFIs must come through the design firm's designated project coordinator. None should be sent directly from the contractor to the design firm's consultants.
- Answers to RFIs must be concise and unambiguous. The point is to make answers so clear that they can be understood by all parties and not misinterpreted. A good test is: Could a jury of laypeople understand the answer to the RFI?
- Do not inadvertently change the scope of work in an RFI. To avoid this, include a statement on the RFI form that says that the RFI is not a change order and if the contractor perceives the answer to be a change in scope, the contractor

shall notify the design professional. On other occasions, the design professional realizes that the answer will change the scope of work. In these instances, the design professional should acknowledge the scope change in the response and say that a change order request will be issued.

- Always run RFIs through a quality control check. A second set of eyes often sees things that are unclear or that could potentially cause confusion. The quality control check gives the design firm a chance to correct inaccurate or ambiguous answers that could cause problems.

Writing Change Orders

Throughout construction, the answers to some contractor RFIs lead to changes in the contractor's scope of work. In addition, owners sometimes want to make changes. When this happens, the design professional prepares a request for quotation (RFQ) or proposal request (PR), which is issued to the contractor, requesting that the contractor prepare and submit a price for the change in scope. As in the case of RFIs, a standardized RFQ form is the best way to issue and administer RFQs. Changes in scope can involve cost, time, or both. The contractor response to the RFQ includes the contractor's costs, markup of subcontractors, profit, additional project-related overhead costs such as additional bond coverage, and supervision. It also includes a request for additional time to do the additional work, if warranted. Like RFIs, the design firm's project coordinator should keep a log of RFQs. At a minimum, the log should keep a record of the RFQ number, its title, date issued, the contractor's quote, the date of the quote, when a contract change order is issued, and the amount of the change order.

The following are some risk management guidelines for RFQs:

- Consult with the owner before issuing any RFQ. It is the owner's money, after all.
- RFQs should be issued to the general contractor only, never to subcontractors. It is the contractor's responsibility to contact the appropriate subcontractors and get their input to the RFQ.
- Review the contractor's response to RFQs carefully. Many owners require the design professional to prepare a written estimate for the RFQ before it is issued to the contractor. Compare the contractor's quote to the estimate for reasonableness, keeping in mind that the quote is a sole source quote and consequently it is not competitive. Look for the time extension quotation and determine if the additional time seems warranted.
- Issue RFQs punctually—if the RFQ is a result of an RFI then the chances are good that it is time sensitive. Issue the RFQ quickly to avoid delaying the contractor.
- Run all RFQs through the design firm's quality control program before issuing them.

Review Contractors' Requests for Pay

Design professionals are often asked to review contractor's requests for pay. In order to know if the pay request accurately reflects the work completed the design professional needs a breakdown of the contractor's contract price into manageable work elements. Early in the project, prior to the first request for pay, the contractor should be asked for a breakdown of the project into materials, systems, and trades with associated costs. This breakdown is reviewed and, if agreed to by the design professional and owner, it becomes the basis for all requests for pay. The breakdown must provide sufficient detail so that when the design professional visits the job site, he or she can determine the percentage of completion of the work, line item by line item. For example, a typical line item might be "reinforcing steel for the slab-on-grade—$30,000." If the contractor requests $15,000, then approximately 50 percent of the steel should be installed. The design professional goes through all line items of the request for pay in a like manner, with each line item judged on its own merits. The overall pay request is the summation of all the line items.

Many projects have retentions. The purpose of the retention is to protect the owner from overpaying and to encourage the contractor to keep working, a bit like dangling a carrot in front of the contractor to keep him or her moving forward. An additional purpose of the retention is to reserve enough money to finish the project should the contractor fail to perform. Retention of 10 percent is very common, meaning that the owner will withhold 10 percent of the contractor's request for pay.

Occasionally owner/design professional agreements include a clause that requires the design professional to certify that the contractor's request for pay is accurate. The clause might read:

> The design professional shall review the contractor's request for pay and shall certify that the amount requested accurately represents the proportion of work completed.(57)

This clause creates additional risk and should be deleted or revised during contract negotiations. If the owner will not accept its deletion, the following modification reduces risk:

> The design professional shall review the contractor's request for pay and render his or her opinion whether the amount requested accurately represents the proportion of work completed.(58)

The following are some risk management rules to follow when reviewing contractors' requests for pay:

- Come to an agreement among the contractor, owner, and design professional regarding the contractor's breakdown of costs prior to the first request for pay.
- Do not overpay contractors. This is particularly important near the end of a project. If a contractor has been paid 99 percent of a particular line item, the contractor has little incentive to finish that line item for the additional 1 percent pay.
- Withhold enough money from pay requests for record submittals. At the end of many projects, contractors must submit as-built drawings. In addition, they must submit operation and maintenance manuals and sometimes provide specified training sessions for the owner's personnel. There are costs associated with this.
- Only pay for materials that are incorporated into the work, not simply stored on site. If agreeable to the owner, it is acceptable to pay for materials stored in a bonded warehouse.
- Review the progress of the contractor's as-built drawings at every pay request.

Sometimes contractors are lax in keeping their as-built drawings (as-builts) up-to-date. At the end of the project, they try to remember what they did, adversely affecting the accuracy of their as-builts. By reviewing them at every request for pay, the contractor knows that the design professional's recommendation of the pay request hinges upon keeping the as-builts up-to-date, and consequently he or she is more likely to do so. Often it is a good idea to include a line item on the contractor's breakdown for record submittals. An alternative is never to pay 100 percent for line items of work that involve as-builts or the submission of record submittals until they are made.

Project Closeout

When a project nears completion, everyone is anxious to finish. Nevertheless, there are still a number of things that must be done. Some contractors are notorious for not completing projects, getting them 98–99 percent done and then moving on, hoping that the owner and design professional will conclude that close is good enough. For design professionals, closeout activities include preparing punch lists of items that need to be completed, issuing certificates of substantial completion, confirming that the contractors' as-built drawings and other record submittals are made, and preparing record drawings based on the contractors' as-builts.

Occasionally owner/design professional agreements include a clause requiring the design professional to certify that the project was constructed correctly. For example, the clause may read:

> Upon completion of construction, the design professional shall certify that the project is constructed in strict accordance with the contract documents and that it fully complies will all building codes, ordinances, and regulations.[(59)]

It is not practical or wise for design professionals to make this certification. During construction, design professionals generally only make periodic observations, not detailed inspections. Consequently, they cannot know if construction complies with all requirements. Design professionals should only certify what they know to be true. Clauses such as the previous one create additional risk and should be deleted or revised during contract negotiations. If the owner will not accept deleting the clause altogether, the following modifications reduce risk:

- Substitute the words "render an opinion whether" for the words "certify that."
- Delete the words "strict" and "fully" and substitute in their place the words "general" and "generally."

The following are two risk management guidelines for closing out a project:

- Prepare a thorough punch list. Make sure all disciplines participate. Contractors frequently expect all of their missing requirements to be on the punch list. Consequently, augmenting a punch list later with additional items can lead to conflicts. To avoid disputes, be as complete and concise as possible.
- Confirm that the contractor's as-builts and record submittals are submitted.

Design professionals do not prepare as-built drawings, they prepare record drawings. They base drawings on documentation provided by the contractor. These drawings are a permanent record of the contract and changes based on the best information made available by the contractor. Thus, the design professional may have to educate the owner regarding the purpose and accuracy of record drawings.

RISK AND DESIGN/BUILD

The design professions are a roller coaster ride; but if a design professional is really looking for thrills, then design/build fills the bill. In a traditional arrangement, the design professional works directly for the owner and bases the design on the owner's program. The project is designed, refined, and construction documents prepared. Contractors then bid the work and the owner enters into a separate contract with a constructor. During construction, the design firm provides construction support services and serves as the owner's agent or representative, watching out for the owner's best interest.

The design/build process changes this relationship. In design/build, the design professional and the constructor are either partners or the same entity. Often the design professional plays the role of consultant to the constructor. This changes the loyalty and the legal relationship between the design professional and the owner.

The design professional is supposed to watch out for the design/build team's best interest and has a legal obligation to do so. This has an impact on the functionality of design, the choice of building materials, the choice of finishes, and the quality of detailing.

For the design professionals to reduce their risk on design/build projects, the following guidelines are helpful:

- Work only on project types for which design/build makes sense. Unique projects are not good candidates for design/build. Repetitive or simple projects are better choices.
- Avoid design/build projects where the building program is unclear or filled with variables.
- Review the design/build agreement carefully. Make sure that what the design professional is promising is covered by his or her liability insurance policy. Make sure that the legal arrangement with the constructor is not somehow increasing the design professional's standard of care.
- Watch out for conflicts of interest. When considering substitutions, it should be clear to the design professional that the constructor is suggesting the substitution to reduce costs.
- Explain the design/build working relationship to the owner.
- Understand the design/build contracting laws in the state where the project is to be constructed. Some states do not allow design/build for publicly funded projects. Others do not allow design professionals to be employees of design/build companies.

CHAPTER SUMMARY

Common construction-phase design services include:

- Submittal reviews
- Construction procurement services

- Construction observation
- Responding to RFIs
- Writing change orders
- Review contractors' requests for pay
- Project closeout

To reduce risk during the review of submittals, design firms should establish standard submittal review procedures that

- make it clear that the project manager is responsible for submittal shop drawing review;
- require a submittal coordinator for every project during construction;
- require a shop drawing log for tracking the status of all submittals;
- require all PMs to use a standardized shop drawing transmittal letter;
- establish filing procedures for submittals;
- establish submittals conformance procedures; and
- require all submittals to run through a quality control review.

Design professionals frequently assist owners during the bid and award phases of projects by providing project procurement services. To reduce risk in providing these services understand the differences among the main project delivery systems:

- Negotiated sole source
- Bid by prequalified contractors
- Generally advertised competitive bidding
- Construction management
- Construction management at risk
- Design/build

To reduce risk when responding to contractor's requests for information (RFIs), develop office procedures for the standardized review of RFIs:

- Prepare and keep a log of RFIs, tracking their status.
- Respond to all RFIs in writing.
- All communication should be through the general contractor, not directly with subcontractors.
- Answers to RFIs must be concise and unambiguous.
- Do not inadvertently change the scope of work in an RFI. If the design professional needs to change the scope of work, tell the contractor that a request for quotation will be issued.
- Always run RFIs through the firm's standard quality control program.

Answers to RFIs occasionally require modifications to the scope of work or the owner may request a change. When this happens, design professionals issue requests for quotation (RFQs) to contractors. Some rules to control risk when issuing RFQs are:

- Always consult with the owner before issuing a request for quotation (RFQ).
- RFQs should always be directed to the contractor, not subcontractors.
- Issue RFQs punctually, particularly if they are generated by a contractor's RFI.
- Review contractor's RFQ quotations carefully.
- Always run RFQs through the firm's quality control program before issuing.

Design professionals are often asked by owners to review contractor's requests for pay and to recommend payment. To reduce risk:

- Request a breakdown of the contract price into manageable work elements by materials, systems, or trades.
- Never overpay the contractor.
- Only pay for materials incorporated into the project. If it's acceptable to the owner, it is okay to pay for materials stored in a bonded warehouse.
- Do not pay 100 percent on line items that require record submittals until the record submittals are made.
- Review the contractor's as-builts at every request for pay to see if the contractor is keeping them up-to-date.

Owners frequently ask design professionals to assist in closing projects. Project closeout services include preparation of a punch list, issuing beneficial occupancy certificates, and receiving and reviewing the contractor's record submittals. Owners also ask design professionals to prepare record drawings.

To reduce risk when working on design/build projects consider the following guidelines:

- Work only on project types that make sense for design/build.
- Avoid design/build projects where the project program requirements are unclear or have many variables.
- Review the design/build agreement carefully.
- Make sure that design/build services are covered by the design professional's liability insurance policy.
- Watch out for conflicts of interest.
- Explain the design/build working relationship to the owner.
- Understand the design/build contracting laws in the state where the project is to be constructed.

Chapter

9

RISK MANAGEMENT RULES OF THUMB

One day, a man walked into a men's clothing store. He found a pair of pants he liked, took them off the rack, and held them up to his chin. He wrapped the waistband around his neck. "I'll take these," he said to the store clerk. "But don't you want to try them on first," asked the clerk, "to see if they fit?" "I already did," said the man.

How did the man know the pants would fit? He applied a centuries-old procedure for approximately calculating the size of his waist called the Tailors' Rule of Thumb. It states that a man's waist is twice the diameter of his neck. Because the waistband was folded in half, he knew that if the ends of the waistband just touched at the back of his neck, then they would fit. It's a quick and practical decision-making tool that yields approximately correct results.

Rules of thumb have many practical uses for design professionals. They are problem-solving shortcuts based on past experiences. Through trial and error, design professionals can determine which rules of thumb work for them. Following are 10 rules of thumb related to both the profession and business of design.

The rules of thumb are based on my personal experience and other surveys of design firms. Like all rules of thumb, they are not exact, but for design professionals that find the rules relevant to their needs, the rules yield practical, easily applied, approximate decision-making guidance.

RULE OF THUMB NO. 1: KNOW THE TERRITORY

On many different occasions I have participated in prequalification interviews for general contractors. A long time ago, an owner's representative and I interviewed staff from a large general contracting firm for a project located outside of the contractor's normal geographic territory. Because the contractor was "hungry" and the contractor had considerable experience with these types of projects, the contractor decided to go after the project and participate in the preselection process. To improve its chances of winning, the contracting company teamed with a smaller contractor from the local area that self-performed concrete work and had recently successfully completed a project for the owner.

During the interview, the vice president of the large contracting company explained why his company was partnering with the local contractor. He said he had learned the hard way "that you have to know the territory." His comment reminded me of something I heard over and over as a child. My parents' favorite musical was *The Music Man* and we listened to it many times. At the beginning of the album, the anvil salesman explains the key to success selling anvils: "You've gotta know the territory." What applies to contractors and anvil salesmen applies equally well to design professionals.

The word *territory* should not be taken literally. Knowing the territory means knowing yourself, knowing the project and client types you serve best, and knowing how you compare to specific competitors and, in general, how you compare to what is normal or average for the industry.

RULE OF THUMB NO. 2: THE 3.0 MULTIPLIER RULE OF THUMB

There are different multipliers with different meanings used by design professionals. Three common ones are: net labor (or net) multiplier, break-even multiplier, and pricing target multiplier.

The net labor (or net) multiplier is calculated by dividing net revenue by the cost of direct labor. This yields a multiplier that tells the design firm how many dollars of revenue it generates for every dollar of labor it spends. For the calculation to work, revenue generated to cover the costs of consultants and project-related expenses are pass-through expenses and must be deducted from the total revenue—hence the term *net.*

The break-even multiplier is equal to the firm's total indirect expenses plus total labor expenses divided by total labor salaries. The resulting ratio equals how many dollars of net revenue are needed for every dollar of labor spent in order to break even (meaning to neither lose nor make money).

The pricing target multiplier is the dollar amount that a design firm wants to charge for every dollar of labor spent. Understandably, the amount that a firm wants to charge is greater than its break-even multiplier. The pricing target multiplier includes profit. Surprisingly, the average pricing target multiplier has changed little for decades. It is 3.0. Therefore, for example, if an employee makes $30 per hour, a 3.0 pricing target multiplier says that the design firm targets a billing rate of $90 per hour for the employee, which includes profit. The 3.0 multiplier is so common it is a commonly used rule of thumb. Not only design firms use it, but owners do also. From an owner's perspective, firms with pricing multipliers over 3.0 are higher-priced firms and these with multipliers less than 3.0 are low-priced firms.

It is interesting to note how long this 3.0 multiplier has been around. According to PSMJ Resources, Inc.'s *A/E Financial Performance Survey* the average pricing target multiplier has hovered around 3.0 for over 20 years.[60] However, I recently learned that it has been around longer than that. In doing research for this book, I came across a 1941 book entitled *This Business of Architecture*, by Royal Barry Wills. Wills explained how to calculate a design firm's overhead rate. After explaining how to do it—total indirect expenses divided by total direct labor costs—he wrote: "An elementary observation makes it clear that, assuming an average drafting wage of one dollar per hour, you have to charge $3.00 an hour against the work of each draftsman..."[61] The 3.0 rule was at play over 60 years ago.

The 3.0 rule of thumb can be used to calculate quickly the fee for designing an element of work. If you know how many hours it takes and the average salary of employees, it is relatively simple to estimate the fee by multiplying hours times salary times 3.0.

The rule can also be used to judge how your firm's net labor multiplier compares to the average price target multiplier of 3.0. If it is more than 3.0, your firm is higher priced than average; if it's less, then your firm is less expensive than average. As mentioned, owners also apply this rule. Naturally, the further your firm's multiplier strays in either direction, the less your firm is like others. To give you an idea of the range and how your firm stacks up, PSMJ's 2006 *A/E Financial Performance Survey* found that the average net labor multiplier among the design firms surveyed was 3.02 and the mean was 2.99. The survey found that 75 percent of the firms had multipliers less than 3.29 and only 25 percent had multipliers less than 2.69.[62]

RULE OF THUMB NO. 3: MANAGE SCOPE CREEP OR IT WILL BREAK YOUR BACK

Like the straws that break the camel's back, the incremental addition of small changes in a project's scope of work can lead to project failure. When owners make big changes, they readily admit it, realize that the design professional is entitled to additional compensation, and are willing to pay for the extra services. In addition, owners generally know that big changes affect both the project's construction cost and the project schedule. Frequently, however, they do not make these connections when it comes to incremental small changes. All is well and good, they think, so long as the design professional never says anything.

Design professionals must guard against the compilation of numerous small changes. Called scope creep, small changes may seem innocuous, at first. Add a room here, a door there. However, they continue: move a wall; how about a few more skylights in this room; a bigger conference room and maybe an operable partition to divide it into two smaller conference rooms; and we should have security cameras and more data/communication outlets. The changes go on and on.

If the changes are made during the schematic design, their impact and design costs may be minor. However, clients often continue making changes throughout the project. More than once, I have worked on projects where the client requested changes the day before we planned to print the final construction drawings!

Besides creating cost and schedule problems, scope creep zaps design team enthusiasm and increases the likelihood of coordination problems and errors, which adversely affects the quality of the construction documents. Poor construction documents can lead to serious problems during construction or beyond.

Nevertheless, many project managers are reluctant to manage scope creep. They do not request fee augmentations for the minor changes—indeed, many PMs don't even acknowledge that the owner has made a change at all. Because the costs of making the changes are minor, they think it is not worth bringing them to the client's attention and doing the necessary paperwork to augment the design fee. Requesting a change, they rationalize, may adversely affect the relationship with the client, so they quietly make the changes. Inevitably, the project creeps over budget. The practice is so common that when projects go over budget, upper management almost automatically assumes that owner-initiated design changes are the culprits.

Project managers must control scope creep. There are a few techniques. The most obvious is to bring the affect of the change to the owner's attention the first time it is requested. This tends to reduce the number of changes because the owner knows the PM will ask for additional compensation with every change. The disadvantage is that the owner may resent so many charges. Because many design firms rely on repeat business, being tagged as a firm that charges for every change may not be a wise business decision. More than once I have heard owners complain about other design firms. "So-and-so nickels and dimes us," they say grudgingly. Nickel-and-diming may improve the financial success of one project, but only at the expensive of losing a valuable client.

Here's a better approach: When an owner makes a minor change, agree to do it, but remind the client of the design costs associated with the change. Regardless of the size of the change, always tell the owner of the construction cost or schedule implications. Politely inform the owner that if too many changes are requested, you will eventually have no choice but to charge for them. With this approach, project managers should set up their project design budgets with a contingency for owner changes. With the contingency, the PM knows that a small number of changes won't break the project. The PM keeps track of the number of changes and when the owner eventually requests one too many, instead of breaking the camel's back, the PM requests the additional compensation that the owner has been made aware of and is prepared to accept.

Some sophisticated owners recognize the problem of owner-initiated scope creep. They include a contingency for it in their project design budgets. On a few occasions, I have been asked by owner project managers to recommend a reasonable contingency amount for owner-initiated design changes. The project managers include the contingency amount in their project budget requests to their upper management, thus giving the project budget a "pad" to cover the cost of owner-initiated changes during design.

RULE OF THUMB NO. 4: DON'T PUSH THE PANIC BUTTON

When I was a teenager, I had a sign hanging in my room: "Around here we don't push the panic button, we sit on it!" Some people have the attitude that nothing will turn out okay. Design professionals shouldn't behave as Chicken Little running around frantically like the sky is falling, needlessly worrying everybody. Nevertheless, they sometimes do. An email sent to everybody in a fit of worry or rage about a problem can throw a design office into a tizzy and grind a project to a halt. Instead of working to solve the problem, people become defensive, spending time explaining their actions and justifying their decisions.

Manage risk by controlling your reaction to unpleasant surprises. If something goes wrong, don't respond in the heat of the moment. Instead, calm down and take a deep breath. There is probably a solution, but you will need a cool head to find it. Blowing off steam or taking knee-jerk reactions won't fix the problem. It might actually make things worse and escalate a little problem into a big one. Instead of wasting time placing blame and firing off inflammatory emails, invest the time and energy on the punch list and on resolving the problem.

I recall an email an owner's representative sent to everyone involved on a particular project, including his boss and his boss's boss. The project was near completion, the punch list had been prepared, and the contractor was slowly working on the punch list—way too slowly, as far as the owner's representative was concerned. He got mad, wrote an email that exaggerated the problem, and accused the contractor and design firm of incompetence. He blamed the contractor because he wasn't finished with the punch list, the design firm because it wasn't making the contractor finish, and his boss because he wasn't making the design firm make the contractor finish. A little problem now escalated into a big problem. Instead of spending time and energy on the punch list, time was spent writing more emails—lots and lots of emails.

RULE OF THUMB NO. 5: THE 60 PERCENT RULE OF THUMB

Not everyone in a design firm does "design work." Many years ago, I worked for an architect who once taught a course on professional practice to architectural students. He started each new class of students with a little eye-opening calculation. Only 60 percent of the time in a design office is spent working on projects of any kind, he told the students. The other 40 percent is spent getting work, managing the firm, doing all the paper work, or lollygagging around. Of the 60 percent, 20 percent is spent providing construction support services, leaving 48 percent of the time for design work. Fifteen percent of the 48 percent is spent doing schematic design. He went on to say that in college design studios, few students advance past the schematic design level with their school design work. "Fifteen percent of 48 percent is 7 percent," he explained, "Now, I'm going to teach you about the other 99 percent of architecture you need to know."

The 60 percent that my friend used in his calculation is alternatively called the "utilization rate" or "chargeability rate." It is calculated by dividing the hours spent working on projects by the total hours spent. For example, if 20 employees spend 472 hours charging to various projects in a 40-hour work week, the utilization rate is 472 hours divided by 800 total hours spent, which equals 59 percent. The other 328 hours were spent in a variety of ways: marketing, office management, vacation, sick leave, professional development, and so on. The utilization rate is used to measure a firm's overall efficiency. In this case, efficiency means how easy it is to invoice and recover the money spent on overhead and salaries. Clearly the easiest and fastest way to recover the money would be to invoice clients for every hour spent. However, this would require a utilization rate of 100 percent, which is not practical.

PSMJ Resources, Inc.'s 2006 survey found that utilization rates vary between design firms by firm size, types of services provided, and client and project type. In general, smaller firms with staff sizes of 1 to 20 experience the greatest deviation. According to the survey, the average utilization rate for all firms is 60.6 percent and the median is 59.8 percent. Twenty-five percent have utilization rates below 55.8 percent and another 25 percent have utilization rates higher than 64.5 percent.[(63)] The survey helps substantiate the validity of the 60 percent utilization rate rule of thumb.

The 60 percent rule can help a design firm determine how it compares to competitors. It can also suggest ways of improving profitability. If a firm has poor profitability and its utilization rate is in the mid-50s or lower, a wise business plan goal for the coming year might be to increase the utilization rate by a percent or two, maybe more. For example, using the 59 percent utilization rate for the 20-person office previously described, it is possible to calculate what a 1 percent increase in utilization would mean to the firm's bottom line. Assuming an average billing rate of $100 per hour, 472 hours multiplied by $100 per hour equals $47,200 worth of billing per week. A 1 percent increase in utilization yields 480 hours per week, which results in $48,000 (480 × $100) per week, an increase of $800. Multiplying this by 52 weeks yields $41,600 of additional profit per year, a tidy sum.

By the way, it is virtually impossible to run at utilization rates above 86 percent for sustained periods of time. This is because vacation, sick leave, holidays, and jury duty account for approximately 14 percent of a design firm's yearly labor hours.[64] While it is possible to run higher than 86 percent for short periods of time when the office is very busy, over the course of a year, the utilization averages out.

RULE OF THUMB NO. 6: PROFITABILITY HAS TWO MAIN RUDDERS

The previous rule about utilization rate might leave the impression that it takes one rudder to steer the design firm. Pick a course—say a 62 percent utilization rate—and check it weekly to make sure that the firm is running at it. Accept a lower rate during the holidays, and there you have it: improved profitability! No, it's not quite that easy. Alas, unlike a ship, a design firm has two rudders and they must be controlled simultaneously. Utilization rate is one rudder, and the other is project profitability.

The utilization rate is a measure of efficiency, a comparison of billable time spent to total time spent. It is neither a measure of productivity nor effectiveness. The utilization rate calculation tells the firm nothing about what was accomplished for the time spent on projects, only the percentage of time spent. In the previous rule of thumb, the 472 hours could have been spent on just one project that had a budget of only 200 hours left. The project would now be 272 hours over budget and drowning in red ink, but the utilization rate would never reveal this fact.

Projects must run on budget. For this to happen consistently, projects must have good project managers and work plans. They must be managed effectively and their progress monitored. When they stray off course or fall behind, corrections need to be made. Design firms need to be cognizant of project performance and budgets. Regular project reviews help. Design firms must watch and control the utilization rate and the projects because they are the two main rudders that control profitably.

I once worked for a large A/E firm that was bought by an even bigger publicly traded E/A firm. Shortly after the purchase, the acquiring E/A firm began playing with the rudders. It changed the way the A/E firm's staff typically charged projects. Under the new rules, office managers and bookkeepers charged more time to projects, with a resulting increase in utilization rate. The problem was that this increased the amount of time, and consequently costs, charged to projects. Soon many PMs' projects were in financial trouble. Many support staff functions that had been treated as overhead were now appearing as charges against projects. There was basically no difference between the effectiveness of the two firms. The difference was simply a matter of where costs were captured. In short, the acquired firm charged fewer hours at higher hourly rates. The acquiring firm charged more hours at lower hourly rates. The overall fees that the two firms needed to do project-related work were the same. The difference was only in the way the two firms manipulated the rudders to control the ship. Changing the rules midcruise made the PMs from the acquired firm look bad, because their projects began running over budget. The higher utilization rate made the acquiring firm look good in the eyes of stock analysts, because utilization rate is seen as a measure of efficiency, in purely financial terms. Because the acquiring firm was publicly traded, higher efficiency was a good thing.

A few months after the acquisition, I attended a meeting where the chief financial officer of the E/A firm explained how the publicly traded company calculated its overhead and utilization rate and how the stockbrokers that watched the industry liked the firm's low overhead rate and high utilization rate. I asked the

CFO if he thought the stockbrokers understood the details of how design firms determine, manage, and massage their overhead and utilization numbers. He answered, "Probably not."

RULE OF THUMB NO. 7: SOMEBODY IS KEEPING SCORE

Many years ago I was involved in arbitration. The contractor filed a claim against the owner alleging that the owner caused financial damage to the contractor. One allegation was that the contract drawings were unclear and had too many mistakes. During the arbitration, the contractor had a claims management consultant testify. During the testimony, it became clear the contractor had hired him at the beginning of the project. He had been reviewing the design firm's responses to RFIs and RFQs, the number of RFIs, RFQs, and change orders, the minutes of job site meetings, shop drawings reviews, the timeliness of the reviews, and so on. His job from the outset of the project was to document anything that could be used later, if need be, to illustrate a pattern of negligence on the part of the owner and design professional.

It is reasonable to assume that everything you do, say, write, or draw may be analyzed by someone willing to use it against you. Beware: Somebody is keeping score.

RULE OF THUMB NO. 8: DRAWINGS ARE NEVER 100 PERCENT COMPLETE

According to the ancient Greek philosopher Zeno, the hare in Aesop's fable can never catch the tortoise. In the fable, the cocky hare is so certain that he can beat the tortoise that he gives the tortoise a head start. However, to catch the tortoise, the hare must first go 1/2 the distance that separates him and the tortoise. Once he has done that, he must then go 1/2 the remaining distance. He always gets closer but he never really gets there. The same is true for construction drawings. They are never 100-percent complete. Incidentally, there isn't even a standard definition for "100 percent complete." There is always something more that could be shown, detailed, or tweaked.

Because it is not possible to be 100 percent complete, and because there is no definition for it anyway, avoid stamping, printing, or otherwise affixing on drawings phrases like "100 percent Complete Drawings," or "100 percent Drawings." It implies a state of perfection that is impossible to achieve. It also suggests that the drawings can be used for construction.

Frequently owners have submittal requirements with names like 35 percent submittal, 75 percent submittal, and 100 percent submittal. The 100 percent submittal does not mean that the drawings are ready or suitable for construction. The purpose of the 100-percent submittal is to give owners one last chance to review and comment on the drawings before they are issued for bid or construction. After the design professional picks up the owner's final comments and finishes the final quality control check, the drawings are as ready as they are ever going to be. They can be submitted to the building department, issued for bid, or given to a contractor for construction. The final set that the contractor builds from should be stamped or printed with the phrase "Issued for Construction" or "Construction Set." For those owners who insist on calling their final submittal the "100 Percent Submittal," consider stamping or affixing the following phrase to the drawings: "100 Percent Submittal—Not for Construction."

RULE OF THUMB NO. 9: THE HURRIEDER I GO THE BEHINDER I GET

There is a logical progression for every design project. It takes a certain amount of time to perform the steps. If the PM or the owner is in a big hurry, it is very easy to upset the logical order of design tasks or to leave tasks out. This frequently leads to mistakes or omissions and increases risk.

Another mistake that design professionals in a hurry make is to start work without a contract. They are anxious to get started when staff is available and enthusiasm is high. They begin designing only to find that somehow fee negotiations broke down or the owner canceled the project at the last minute.

Frequently design professionals start the next phase of the project before receiving comments from the owner or before receiving approval from the owner to start the next phase. Some contracts require owners to issue notices to proceed for subsequent phases. Without the owner's comments or approval to proceed, the design firm has increased risk. More rework may be needed because design work proceeded without feedback from the owner, or worse, the owner may decide not to go forward with the project and instead of issuing a notice to proceed, shuts the project down.

Design professionals are frequently in a hurry to start construction documents before the design is fully figured out. Even though there are still major design issues to resolve, staff is busily producing construction documents that will have to be changed once the design is worked out. Minimize this risk by completing the design phase before beginning production of construction documents.

RULE OF THUMB NO. 10: DOCUMENT, DOCUMENT, DOCUMENT

Your memory isn't as good as you think. Even if it is, people are more likely to trust you if you can produce a written record to confirm that your memory is correct.

We have heard people testify before Congress saying that they "don't recall." "I can't recall" isn't just their excuse for a bad memory; it is actually their defense. A design professional's best defense is producing documentation that supports what he or she says. In rule of thumb 7, we learned that somebody is always keeping score. Because they are, make them give you a high score. Some general rules for documents are:

- Think before sending emails
- Don't vent in a diary (your work diary is a document that can be subpoenaed)
- Document when owners say they like the design
- Document when they approve moving to the next phase
- Document design decisions
- Document when owners request changes
- Document meetings
- Document, document, document

RISK MANAGEMENT CHECKLIST

Following is a chapter-by-chapter summary and checklist. It takes the risk management topics discussed throughout the book and consolidates them for easy reference.

CHAPTER 1 PRINCIPLES OF RISK MANAGEMENT

There are a few fundamental principles used to make decisions to control outcomes when the future is unknown. These principles are the basic decision-making tools of risk management. Employing them simultaneously increases the chances of success while reducing the likelihood of failure. The principles are:

- ✓ Some outcomes are more likely than others. Take actions to make desirable outcomes the likely outcomes.
- ✓ Fractions (percentages) can be used to express the probabilities of different outcomes. Use them to evaluate the probable successes of alternative decisions and actions.

- ✓ Risks and rewards are not always proportional. Avoid taking big risks for little reward.
- ✓ The whole can be estimated from the parts. Generalities about the whole can be gleaned from looking at samples or parts of the whole.
- ✓ Facts and utility are not the same thing. Utility is subjective and it is the a measure of usefulness.
- ✓ Outcomes are predictable within a range.
- ✓ Managing the vital few is more effective than managing the trivial many. Employ the 80/20 rule.
- ✓ Risk management is a rational process involving all three logic tools of inference: deduction, induction, and abduction.
- ✓ Abduction is the most commonly used inference tool in risk management. Abduction is pragmatic educated guessing about what is likely to happen based on relevant past experience.

The rational decision-making process includes the following steps:

- ✓ Establish the premises or factors that bear on the decision.
- ✓ Identify possible alternative decisions that satisfy the premises or factors.
- ✓ Analyze the merits of the alternative decisions, considering all relevant data, making sure to consider the risks associated with competing alternatives.
- ✓ Determine the likelihood of the risks occurring.
- ✓ Choose the decision that logically offers the best solution considering the relevant data and risks involved, knowing that the decision contains some uncertainty because decisions about the future always do.

CHAPTER 2
THE RISK MANAGEMENT PROCESS

Applying the rational decision-making process described in Chapter 1 to risk management yields the following six steps of the risk management process.

1. Acknowledge risk: Recognize that there are many risks that affect the practice and business of design and that these risks can be controlled by a comprehensive risk management plan.
2. Determine the risks: There are many that adversely affect success. They can be grouped into 15 risk-source categories (see the list after this one).
3. Determine the probability: The chances of various risks occurring vary depending upon particular circumstances. Probabilities can be determined quantitatively or qualitatively.
4. Determine the severity of the impact: Quantitatively or qualitatively predict the consequences should certain threats materialize.
5. Develop procedures and solutions to manage risks: There are six basic strategies used to manage risk as described later in this section.
6. Implement those solutions or procedures that manage those risks that have the highest overall negative impact on the desired outcome. Develop metrics for measuring results and monitor the effectiveness of the implemented solutions and procedures.

It is important to understand the sources of risk that affect the design professions. They can be grouped into 15 risk-source categories:

1. Project management
2. Technical performance

3. Organizational
4. Business development
5. Financial
6. Contractual
7. Insurance
8. Quality
9. Employee
10. Consultants
11. Client and project types
12. Technological and work product standards
13. Bid and award
14. Construction phase
15. External

There are six fundamental strategies for the proactive management of risk and one for managing the aftermath or effects. They are:

1. Prevention—reduces risk by reducing the number of uncertainties and/or threats.
2. Mitigation—lessens the impact of the uncertainty or threat should it occur.
3. Transference—passes or transfers the risk to some other party, making it responsible for the consequences of the risk.
4. Contingency planning—the advanced preparation of an alternative plan for dealing with specific uncertainties or threats should they occur.
5. Assumption—the identification and analysis of the impact of specific risks and the acceptance of their potential impact.
6. Recovery planning—reactive planning that deals with managing the effects or outcome of a risk that has materialized.

Overall risk is determined by multiplying the probability of the risk times the impact of the risk. Risk assessment tables and risk assessment matrixes can be used to compare various overall risks, thus assisting the risk manager in determining which risks require plans for managing them.

CHAPTER 3 COMPREHENSIVE RISK MANAGEMENT

Design professionals need an all-encompassing or comprehensive view of risk management. Comprehensive risk management is proactive and engages everyone in the design firm in managing risk. Risk can be managed best by developing a variety of interrelated plans that together address the risks that affect the practice of design. The plans are:

- ✓ Business plan—addresses risks that are organizational in nature.
- ✓ Employee relations and management plan—addresses employee-related sources of risk.
- ✓ Office procedures plan—addresses risks associated with technical performance and production of the work product.
- ✓ Business development plan—addresses the risks associated with marketing design services.
- ✓ Project management plan—addresses risks associated with managing projects and their scope, schedule, and budgets.
- ✓ Claims response plan—addresses the risks associated with the prevention and response to legal and financial claims.
- ✓ Quality assurance plan—addresses risks associated with management of quality of design work and the production of contract documents. It also addresses the quality of other work products such as reports, addenda, responses to RFIs, and preparation of contractor RFQs.

CHAPTER 4
RISK AND BUSINESS DEVELOPMENT

Design professionals need effective and efficient business development plans. Comprehensive business development plans increase the likelihood of winning more work, while simultaneously reducing the risk of failure. In developing a business plan, test it with the following list of questions:

- ✓ Is the plan procedural? Is it composed of specific steps that have been shown to work and that can be followed by all involved in business development?
- ✓ Are the procedures repeatable? Can they be followed time and time again with success?
- ✓ Are the procedures learnable? Can others learn the process, follow it, and also be successful?
- ✓ Is the plan disciplined? Does the plan discourage BD staff from chasing anything that comes along? Will the BD staff concentrate its efforts on pursuing only those projects/clients that are targeted in the plan?
- ✓ Does the plan identify desirable or targeted market sectors?
- ✓ Is the plan focused on project type, client type, or both?
- ✓ Does everyone involved in business development know their roles and expectations?
- ✓ Does the plan have goals? Does every BDer have sales goals, usually expressed as dollar's worth of sales?
- ✓ Does the overall BD effort have an efficiency goal, usually expressed as the ratio of BD costs to sales?
- ✓ Can results be monitored? Have metrics been established to track results for comparison against personal BD and overall business development goals?
- ✓ Is the BD plan consistent with the firm's business plan?

To develop a successful business development plan, honest answers to the following questions are essential:

- ✓ What are the firm's strengths and weaknesses?
- ✓ Is the BD effort set up with seller/doers, sellers-and-doers, or a combination of both?
- ✓ Does the firm want to chase project types, client types, or a combination of the two?
- ✓ What size projects does the firm want to chase?
- ✓ What kind of clients is the firm interested in: commodity or value-added; new or repeat; public or private?

Business development plans usually include the following parts:

- ✓ Business development mission statement—a summary of the plan's purpose and desires.
- ✓ Organizational structure—how the BD effort is organized, including an explanation of its reporting structure, and names of key individuals. Often included is an organization chart.
- ✓ Marketing goals—includes individual sales goals and overall goals.
- ✓ Marketing approach and procedures.

It is useful to develop a standardized procedure for deciding which projects/clients to pursue. Often summarized in a Go/No Go Form, the following questions should be considered when deciding whether to pursue a particular project:

- ✓ Has the firm been tracking the project?
- ✓ Is the project pre-sold? If it's pre-sold to the design firm then it's good news; pre-sold to another design firm is not good news.
- ✓ Is the project for an existing client?

- ✓ Do we understand the client's needs and expectations?
- ✓ Are we qualified for the project?
- ✓ Is it within the market areas stated in our business development plan?
- ✓ What is the selection process?
- ✓ If the selection process involves a selection panel, do we know individuals on the selection panel?
- ✓ What are the likely marketing costs?
- ✓ What is the likely design fee?
- ✓ Do we have qualified staff to perform the work?
- ✓ Do we have a qualified project manager available?
- ✓ Do we have marketing staff available to pursue the project or are we too busy right now?

There are risks associated with writing proposals. Risk increases when the firm oversells its qualifications or states that it can provide services it cannot. Avoid increasing risk by embellishing the firm's capabilities and creating owner false expectations with phrases and words such as the following:

- ✓ "Our firm will prepare complete and detailed drawings and specifications." Drawings and specifications are almost never complete or fully detailed.
- ✓ "We will ensure that…" or "we insure that…" Design firms should not ensure or insure performance or outcome.
- ✓ "We guarantee that…" or "we warrant that…" Design firms should not guarantee or warrant performance or outcome.
- ✓ "We have highly trained professionals," which implies that the firm's design professionals will provide services that are better than the normal standard of care.
- ✓ "We will attend all necessary meetings." This means that the firm agrees to attend an unlimited number of meetings for the stipulated fee.

- ✓ "We will conduct a field survey and gather all existing information." It is not possible to know that *all* information has been attained.
- ✓ Avoid words that create expectations of absoluteness such as "always," "must," "best," and "maximize."
- ✓ Avoid their opposites, too—"never," "worst," "minimize," and "zero"—because these words also create expectations of absoluteness.

CHAPTER 5
RISK AND PROJECT MANAGEMENT

Successful design firms are successful because their projects are successful. Successful projects depend upon the skills of the project manager. To avoid the risk of choosing the wrong project manager, the proposed project manager should have

- ✓ a positive attitude toward authority.
- ✓ a high motivation to lead.
- ✓ good organizational skills.
- ✓ design and technical competency in the project type.
- ✓ a generalist's attitude.
- ✓ the ability to make decisions with incomplete information.
- ✓ strong professional and ethical behavior.
- ✓ good communications skills.
- ✓ the ability to motivate design teams.
- ✓ a risk manager's attitude.

To reduce the risk of project failure, consider the following before starting the project:

- ✓ Is the project impossible due to technological or knowledge limitations? (Projects can fail due to technological and/or knowledge limitations.)
- ✓ Are one or more project objectives impossible to achieve? (Projects can fail if project objectives are unrealistic.)
- ✓ Is it impossible for the project to succeed because of the way it was sold? (For example, if the design fee is too low it may not be possible to complete the project on budget, regardless of the design team's efforts.)
- ✓ Did the project manager and design team participate in determining the fee?
- ✓ Does the project have any unusual scope, budget, or schedule requirements?
- ✓ Has the project team worked together before?
- ✓ Is there a project designer available?
- ✓ What is the project's proposed construction contracting method?
- ✓ Is the project politically sensitive?
- ✓ Does the owner have realistic expectations?
- ✓ Can the firm accept the owner's agreement?
- ✓ Can the firm accept the owner's insurance requirements?
- ✓ Is the execution of the project properly planned? (Projects fail if they are not well planned.)

Some factors that affect project risk should be addressed during the project planning process. Questions to consider that reduce project risk are:

- ✓ Is the scope of work clearly defined?
- ✓ Can the entire project be executed in-house or are consultants necessary?
- ✓ If consultants are necessary, has the firm worked with the consultants before?
- ✓ Is there an adequate design contingency?
- ✓ Are some of the owner's requirements subject to change?

- ✓ Does the project require the design professional to invent something or use new technologies?
- ✓ Is the city, county, or other jurisdiction where the project is located easy to work with?
- ✓ Is the project budget sufficient?
- ✓ Is the owner's construction budget adequate?
- ✓ Is the project schedule realistic?
- ✓ Is there a licensed design professional for every discipline?
- ✓ Is the project work plan thorough and complete? Does it define the project scope so all design team members know what is expected?
- ✓ Are all the project team members identified and their responsibilities defined?
- ✓ Is the project quality control plan defined?
- ✓ Are all project-specific procedures defined?
- ✓ Has the project work breakdown structure (WBS) been determined? Is it in sufficient detail to allow monitoring of progress?
- ✓ Has a written project work plan been distributed to all key team members?

Some factors that affect project risk must be addressed during project execution. Questions that the project manager should consider to reduce project risk during execution are:

- ✓ Is the owner being kept informed of all potential problems?
- ✓ Is the owner involved in making decisions?
- ✓ Are all decisions being documented?
- ✓ Are decisions being made in a timely manner?
- ✓ Are all documents being checked following the firm's quality control program before they leave the office?
- ✓ Before starting construction documents, are all the major decisions made and the design approved by the owner?

- ✓ Is the interdisciplinary work being properly coordinated?
- ✓ Is the project being designed within the design budget?
- ✓ Is the project being designed within the construction budget?
- ✓ Are the project financials being reviewed regularly and progress monitored to confirm that the work is being performed within budget?
- ✓ Is all the work being performed necessary and within the scope of work?

CHAPTER 6
COST, CONSULTANTS, AND CODES

Projects should be run on budget. There are two budgets to consider: the cost of construction and the cost of design. To control the risk of going over the construction budget take the following steps:

- ✓ Determine a realistic construction budget.
- ✓ Break the construction budget down into major components such as by materials and systems (foundation, finishes, HVAC, electrical, etc.) or by discipline (civil, architecture, structural, mechanical, electrical, etc.) and assign budgets.
- ✓ Include a construction cost contingency budget.
- ✓ Require all disciplines to design within their construction cost budgets.
- ✓ If a material, system, or discipline appears to be exceeding its budget, take corrective actions to control the construction cost overruns.
- ✓ If need be, bolster budgets by transferring money from the contingency budget.

To control the risk of going over the design budget take the following steps:

✓ Determine a clear scope of work.
✓ Develop a design budget that makes sense in comparison to the effort required to perform the scope of work.
✓ Break the overall budget down into manageable pieces.
✓ Determine an appropriate design contingency for the project.
✓ Develop a project work plan that respects both the scope of work and the budget.
✓ Control incremental increases in the scope of work—scope creep.
✓ Monitor design progress regularly, comparing progress against both scope and budget.
✓ Adjust the work plan when progress and budget get out of sync.
✓ Keep the client informed of client-initiated changes that affect the scope and budget.

When choosing consultants consider the following:

✓ Work with consultants with which the design firm has good track records.
✓ Using a new consultant is like hiring a new employee. Make sure the new consultant is a good fit.
✓ When using owner-selected or owner-recommended consultants for the first time, interview them and check out their references.
✓ Make sure the consultant meets the owner and the design firm's insurance requirements.
✓ Confirm that the consultant is licensed in the state where the project is to be constructed.

✓ Provide the consultant with the design firm's drafting and QA/QC standards/requirements and require them to follow them.

Avoid the risk of delays in building permitting by starting the code and planning analyses early on in the project. Consider the following:

✓ Design firms should require all PMs to address code analysis/review procedures in the project work plan.
✓ Design firms should establish a list of qualified individuals within the firm to conduct code analyses.
✓ Design firms should establish standardized code review procedures. Consider developing a standardized form or table for code reviews.
✓ Conduct a preliminary code review at the outset of the project. Determine which codes, ordinances, and regulations will bear on the design.
✓ Determine the submission requirements for the regulatory bodies.
✓ Determine the timing and how long it takes for various reviews.
✓ Determine if there is a planning or architectural review board process.
✓ If acceptable to the regulatory bodies, meet with them soon in the development of the design.
✓ Prior to submission to the building department, review the code analysis. Prepare a code summary information sheet for the contract drawings.

CHAPTER 7
PROFESSIONAL LIABILITY AND CONTRACTUAL RISKS

Product liability centers on the principle that manufacturers of products must produce products that are free from defects. Professional liability hinges on the concept that that design professionals are expected to perform in a manner similar to other design professionals practicing under the same circumstances in the same location and legal jurisdiction. Product liability focuses on the outcome—is the product free of defects? Professional liability focuses on the process—how did the design professional perform compared to his/her peers? Design professionals are expected to

- ✓ accomplish the scope of work defined by the agreement with the owner.
- ✓ perform in a manner that is professionally reasonable.
- ✓ do as other design professionals would do under the same circumstances in the same legal jurisdiction.

Design professionals carry professional liability insurance for the following reasons:

- ✓ As a risk management strategy—transference of risk to another party, the insurance underwriter
- ✓ As legal defense against alleged so-called negligent acts, and errors and omissions
- ✓ As financial protection
- ✓ To increase the types of clients and projects that can be pursued because many clients require design professionals to carry professional liability insurance
- ✓ As a requirement of owner/design professional agreements
- ✓ To assist in developing sound risk management practices

Insurance agents and insurance brokers sell professional liability insurance. Insurance agents work directly for insurance companies. Brokers are independent and can represent multiple insurance companies. Questions to consider when choosing an insurance broker or agent are:

- ✓ Does the insurance broker have access to many insurance carriers?
- ✓ Can the insurance company write policies in the firm's geographic areas of practice?
- ✓ Does the broker or agent seem qualified and have a commitment to help the design firm?
- ✓ Is the insurance company an admitted carrier, meaning does it comply with the state rules for guaranteeing funds?
- ✓ Can the broker or agent assist the firm in risk management planning?
- ✓ Can the insurance carrier write additional liability policies if needed, such as project insurance or expanded coverage?
- ✓ Does the policy cover the firm's employees as well as the firm's owners?
- ✓ What are the policies preclusions?
- ✓ What is the deductible?
- ✓ What is the level of design professional participation in handling claims?
- ✓ Are the fees or commissions reasonable compared to the coverage and services provided?

Professional liability insurance policies fall into one of three policy-basis categories:

1. Occurrence—provides coverage even after the policy is cancelled provided it was in effect when the alleged acts or services that caused the claim occurred.

2. Prior acts—provides coverage for acts or services that took place before the insured had insurance or when the insured had insurance with another insurance carrier.
3. Claims-made—provides coverage only when the alleged acts or services and the claim occur while the policy is in effect. Claims-made policies are the most common type of professional liability insurance policies.

Design firms can manage their risk and exposure to claims in the following ways:

- ✓ Determine all of the design firm's insurance needs and attain coverage and limits of coverage consistent with those needs.
- ✓ Develop a productive working relationship with an insurance broker or agent.
- ✓ Avoid working for sue-happy owners.
- ✓ Use standard agreement forms whenever possible.
- ✓ Always be on the lookout for competent staff.
- ✓ Provide risk management training for staff members.
- ✓ Implement a QA/QC program.
- ✓ Avoid agreement clauses that increase risk.

From the point of view of insurance companies, a claim occurs when the following three conditions are met. In actual practice, insurance companies generally agree that a claim has occurred if the first two conditions are met.

- ✓ There must be an injury—physical, financial, or both.
- ✓ There must be an alleged wrongdoing on the part of the design professional.
- ✓ There must be a demand for compensation against the design professional.

If a claim occurs, take the following steps.

- ✓ Assign someone to be the claims manager.
- ✓ Report the claim to the insurance carrier within the time frame stipulated by the policy.
- ✓ Follow all insurance policy requirements and procedures.
- ✓ Maintain confidentiality.
- ✓ Do not admit fault as this could jeopardize insurance coverage.
- ✓ Do not agree to pay money or provide services for claims as this may jeopardize insurance coverage.
- ✓ Do not publicly blame other parties, particularly the firm's consultants.
- ✓ Gather together all relevant documentation.
- ✓ Notify past employees who might have worked on the project if their assistance is needed.
- ✓ Contact consultants who worked on the project who may be affected, because they will have to defend themselves through their own insurance policy.
- ✓ Identify an in-house expert or consider hiring an expert witness because the claimant is likely to hire one.

Risk is reduced if the owner/design professional agreement is properly written. Consider the following risk management rules when reading/writing agreements for design services:

- ✓ Use standardized agreements whenever possible.
- ✓ Understand the provisions of the agreement.
- ✓ Avoid language that increases risk.
- ✓ Avoid unacceptable risks.
- ✓ Use appropriate fee-types for the services provided. The five most common fee-types are: lump sum, time and materials, time and materials with a maximum not to exceed, cost plus fixed fee, and cost plus fixed fee with a guaranteed maximum price.
- ✓ Specify how disputes will be resolved.

Words, phrases, and clauses that increase risk are:

- ✓ Indemnity clauses that require the design firm to indemnify and defend the owner against all claims, not just those due to the design firm's negligence
- ✓ The requirement to guaranty, warrant, or certify things that are beyond the control of the design firm
- ✓ Words such as: insure, ensure, certify, guarantee, optimize maximize, minimize, supervise

Going to court to resolve claims can be very expensive. Consequently many agreements include alternative dispute resolution techniques. The three most common are:

1. Partnering—parties agree to work together to resolve problems in a spirit of cooperation, with the intention of solving problems before they become points of contention of dispute.
2. Arbitration—opposing parties agree to present their disputes to one or more individuals who are empowered to render a decision.
3. Mediation—a voluntary method of resolving disputes whereby opposing parties agree in advance to resolve disputes themselves with the assistance of an independent mediator. This method is the recommended form of conflict resolution in the AIA's *Standard Form of Agreement Between Owner and Architect.*

CHAPTER 8 RISK AND CONSTRUCTION SUPPORT SERVICES

There are six common project delivery systems for construction projects. They are:

1. Negotiated sole source—only one qualified contractor is asked for a proposal.
2. Bid by prequalified contractors—a small group of contractors are selected based on their qualifications and they prepare competitive bids.
3. Open competitive bid—the project is advertised for bids, which are received from all contractors that meet certain legal requirements, and the low bidder is selected.
4. Construction management—a construction management (CM) firm breaks the project down into various bid packages that are competitively bid separately. The owner enters into prime contracts with multiple contractors, which are managed by the CM firm.
5. Construction management at risk—the same scenario as construction management except that CM gives the owner a guaranteed maximum price to deliver the project. Instead of the owner carrying multiple prime contracts, the CM enters into multiple contracts.
6. Design/build—a design/build firm gives an owner a guaranteed maximum price to design and construct the project.

Design professionals provide a number of services during construction. The common ones are: construction procurement services, submittal reviews, construction observation, responses to contractors' requests for information (RFIs), writing of change orders, reviewing contractor's requests for pay, and project closeout.

To reduce risk during submittal reviews consider implementing the following:

- ✓ Establish standard procedures for the review of submittals and require all PMs to follow them.
- ✓ Every project should have a designated submittal coordinator.
- ✓ All shop drawings should be logged in when received. The log should track the date received, submittal number, name of submittal, specification reference, to whom distributed, review action taken, and date returned.
- ✓ All submittals should be reviewed in a timely manner. Many contracts have time-limit requirements for the review of submittals.
- ✓ Prepare a standardized shop drawing transmittal review letter and stamp.
- ✓ Use actions that are consistent with insurance carrier requirements. For example, some may not allow the use of the action "approved."
- ✓ Establish filing procedures for submittals.
- ✓ Establish shop drawing conformance standards—how are redlined review comments accurately transferred to multiple copies of submittals?
- ✓ Prior to construction, provide the contractor with a list of required submittals.
- ✓ Require contractors to review, coordinate, and approve subcontractor submittals before they are submitted to the design professional for review.
- ✓ Require contractors to bring substitutions and deviations to the contract to the design professional's attention in the submittal transmittal letter.
- ✓ Do not review submittals for construction methods, means, procedures, techniques, or sequencing, as these are the sole responsibility of the contractor.
- ✓ Do not review submittals for temporary aspects of the construction process, such as shoring or scaffolding, as these are the contractor's sole responsibility.

- ✓ Do not inadvertently change the scope of work through the shop drawing review process.
- ✓ Require all submittals to run through the firm's standard quality control program before returning to the contractor.

Design professionals observe construction in order to satisfy the owner that the work is being performed in accordance with the contract documents. Design professionals do not direct construction activities. They are not responsible for construction methods, nor are they responsible for construction safety.

To reduce risk in responding to contractors' requests for information (RFIs), consider the following:

- ✓ Establish standard procedures for responding to RFIs and require all PMs to follow them.
- ✓ Develop an RFI log that tracks the RFI number, date received, title, response, and date of response.
- ✓ Require all RFIs to be in writing.
- ✓ All answers to RFIs should be in writing.
- ✓ Only accept RFIs that come from the general contractor, not directly from subcontractors.
- ✓ Answers to RFIs must be concise, unambiguous, and based on the contract requirements, not on what the design professional wished the documents said.
- ✓ Avoid inadvertently changing the scope of work when responding to an RFI. If a scope change is warranted, respond with a request for quotation (see next list).
- ✓ Always run responses to RFIs through the firm's standardized quality control program.

To reduce risk in writing requests for quotations (RFQs) for change orders, consider the following:

- ✓ Consult with the owner before issuing an RFQ and get the owner's concurrence.

- ✓ RFQs should be issued to the general contractor only, not to subcontractors.
- ✓ Issue RFQs punctually.
- ✓ Review contractors' responses to RFQs carefully and promptly. Review for cost and time implications.
- ✓ Issue change orders promptly.
- ✓ Run all RFQs through the firm's standardized quality control program.

To reduce risk in reviewing contractors' requests for payment, consider the following:

- ✓ Ask for a breakdown of the contractor's pay request in sufficient detail to permit assessments of percentages of completion for different portions of the work.
- ✓ Review the status of the contractor's as-built drawings at every request for pay.
- ✓ Do not overpay contractors.
- ✓ Retain enough money for project closeout record submittals.

To reduce risk when working on design/build projects consider the following:

- ✓ Does the project type make sense to design/build?
- ✓ Are the project program requirements clear? Are there too many variables?
- ✓ Has the design/build agreement been carefully reviewed?
- ✓ Does the firm's professional liability insurance policy cover the design/build services? Has this been confirmed?
- ✓ Are there any real or apparent conflicts of interest?
- ✓ Have the design/build working relationship been explained to the owner?
- ✓ Have the design/build contracting laws in the state where the project is to be constructed been reviewed?

Acknowledgments

To properly acknowledge everyone who has helped me understand and appreciate the multifaceted nature of the design profession is impossible. Over the years, more professors, friends, coworkers, architects, and engineers than I can count have contributed to my learning process, teaching me the artistic, technical, legal, managerial, business, promotional, and collaborative aspects of the profession and business of design. Nevertheless, there are a few I want to mention. First, I extend my gratitude to Whitson Cox, FAIA, my first employer, who was willing to hire me right out of school. I want to thank Frank B. Hunt, FAIA, who kept me on when times were tough, and John G. Wells, AIA, who let me tag along to meetings with clients, contractors, and consultants until one day I knew enough to lead design teams on my own. I want to express my appreciation for Richard Kennedy, PE, (1909–1985) and his son David Kennedy, PE. Through example, the two of them taught me the value of personal and professional honesty and integrity, which is paramount to a long career in the design professions. By contrast, many others have shown me the risks in behaving otherwise. I thank my good friend, coworker, and architect Daniel Wright for his sound advice through the years and his patience listening to me vent my frustrations more times than I—and I'm sure he—care to remember.

I wish to thank my son Andrew for helping me with the research, math, and statistics used in the book, and my wife, Persis, for her attentive reading and rereading. Writing often becomes a family affair and so I thank the rest of my family for their support and understanding.

I also thank Victoria Smith of Kaplan Publishing for suggesting that I write a book on the subject of risk management for design professionals, and Joshua Martino, development editor, for his excellent suggestions. Finally, I thank Pamela Brodowsky of International Literary Arts for her help, advice, and encouragement.

Notes

1. Some historians give Robert Mills (1781–1855) the distinction of being the first American architect because he was the first to open an office solely for the practice of architecture in 1810. However, Charles Bulfinch considered himself, called himself, and was thought of as an architect by his contemporaries well over a decade earlier.
2. William H. Pierson, Jr., *American Buildings and Their Architects, Volume 1: The Colonial and Neoclassical Styles* (New York: Oxford University Press, 1970), 241–242.
3. Ibid., 244.
4. Ellen Susan Bulfinch, *The Life and Letters of Charles Bulfinch, Architect with Other Family Papers* (Boston: Houghton, Mifflin and Company, 1896), 94.
5. Ibid., 99.
6. Ibid., 42.
7. Ibid., 100.
8. Vitruvius, *The Ten Books on Architecture,* c. b.c. 30, translated by Morris Hicky Morgan, originally published in 1914 (New York: Dover Publications, 1960), 87–88.
9. Hugh Morrison, *Louis Sullivan: Prophet of Modern Architecture,* originally published in 1935 (New York: W. W. Norton & Company, 1998), 67. For the reader unfamiliar with the "Columbus Egg Stunt," legend has it that to secure the ships he needed to sail around the world Columbus had to prove his worthiness for the impossible mission by doing the impossible—standing an egg on end. As he stood the egg on the table he pressed down firmly, breaking and flattening the shell, thus performing the impossible to the delight and amazement of King Ferdinand and Queen Isabella.
10. Girolamo Cardano, *The Book of My Life,* translated by Jean Stoner (New York, New York Review of Books, 2002), 50.
11. Peter L. Bernstein, *Against the Gods: The Remarkable Story of Risk* (New York: John Wiley & Sons, Inc., 1998), 44.
12. "Famous Gamblers: Chevalier de Mere," http://www.casinos.net/chevalierdemere.htm.
13. Bernstein, Op Cit., 67.
14. Blaise Pascal, *Pensées,* first published in 1669, translated by A. J. Krailsheimer (New York: Penguin Books, 1966), 150.
15. Ibid., 152.
16. Ian Hacking, *The Emergence of Probability: A Philosophical Study of Early Ideas about Probability Induction and Statistical Inference* (New York: Cambridge University Press, 1975), 64.

17. Bernstein, Op Cit., 69.
18. Jacob Bernoulli, *The Art of Conjecturing.* Originally published posthumously in 1713. Translation by Edith Dudley Sylla, *The Art of Conjecturing, Together with Letter to a Friend on Sets in Court Tennis* (Baltimore: The Johns Hopkins University Press, 2006), 317–318.
19. Bernstein, Op Cit., 123.
20. Ibid., 118.
21. G. Simmons, *Calculus Gems* (New York: McGraw Hill, 1992), 136.
22. Bernstein, Op Cit., 103.
23. Ibid., 105.
24. Peirce Edition Project, *The Essential Peirce: Selected Philosophical Writings, Volume 2 1893–1913* (Bloomington and Indianapolis: Indiana University Press 1998), 95.
25. Ibid., 234.
26. Tom Stoppard, *Rosencrantz & Guildenstern Are Dead* (New York: Grove Weidenfeld, 1967), 14.
27. Ibid., 125.
28. Robert Heller, *Making Decisions* (New York: DK Publishing, Inc., 1998), 20.
29. FM Global History, http://www.fmglobal.com/about/facts/facts_history.asp.
30. Roger Hale Newton, *Town & Davis Architects: Pioneers in American Revivalist Architecture 1812–1870 Including a Glimpse of Their Times and Their Contemporaries* (New York: Columbia University Press, 1942), 95.
31. Ibid., 96.
32. Ibid., 311.
33. Ibid.
34. Henry-Russell Hitchcock, *The Architecture of H. H. Richardson and His Times,* first published by the Museum of Modern Art, 1936 (Cambridge, MA: The M.I.T. Press, 1970), 299.
35. James Marston Fitch, *American Building: The Historical Forces That Shaped It,* first published by Houghton Mifflin Company, 1947 (New York: Schocken Books, 1973), 192.
36. More recently, architectural historians have cast doubt on the conclusion that Richardson was a proto-modern architect. Hitchcock, in advancing his thesis, may have been too heavily influenced by his own desire to find a progenitor for modern architecture, a style of architecture that Hitchcock much preferred to traditional architecture. Today Richardson's architecture is considered less revolutionary and more in tune with the style of his time, which was an eclectic mixture of John Ruskin's arts and crafts movement, gothic, picturesque, Victorian, and Beaux Arts classic elements. Richardson did not

write about his work, as many other modern architects have. Consequently his buildings stand, and talk on their own, and what they say can be interpreted differently by different historians. The architectural historian Margaret Henderson Floyd summed it up best in her essay "Inspiration and Synthesis in Richardson's Paine House," in which she wrote that Richardson may have simply "played the game of eclecticism...more ingeniously that his peers." Maureen Meister, editor, *H.H. Richardson: The Architect, His Peers, and Their Era* (Cambridge, MA: The MIT Press, 1999), 39.

37. Peter Piven and Bradford Perkins, *Architect's Essential of Starting a Design Firm* (Hoboken, New Jersey: John Wiley & Sons, Inc., 2003), 17–18.
38. I tried to find the original source for the Richardson legend, but alas, I could not. The best I could do was to find a retelling of it by Piven and Perkins in the book referenced in the previous footnote. The second version, "getting the next job, getting the next job, and getting the next job," I first heard so long ago I can't remember where I heard it. In doing research for this book, I found it again on the Internet. It sounds so much like the three most important things in real estate—location, location, and location—that I suspect it is apocryphal.
39. I thank the architect and skilled business developer Bill Proctor for the funnel analogy.
40. William G. Ramroth, Jr., *Project Management for Design Professionals* (Chicago: Kaplan AEC Education, 2006), 41–53.
41. American Institute of Architects, *The Business of Architecture AIA Firm Survey* (Washington, D.C.: American Institute of Architects, 2007), 48.
42. PJMJ Resources, Inc., *A/E Financial Performance Survey,* 26th ed. 2006 (Newton, MA: PSMJ Resources, Inc., 2006), 25.
43. William H. Pierson, Jr. "The Beauty of a Belief: The Ames Family, Richardson, and Unitarianism," *H.H. Richardson: The Architect, His Peers, and Their Era,* Maureen Meister, editor (Cambridge, MA: The MIT Press, 1999), xxvi.
44. Matthys Levy and Mario Salvadori, *Why Buildings Fall Down: How Structures Fail* (New York: W. W. Norton & Company, 2002), 114.
45. Tax Payers for Common Sense, "Big Dig Billions Over Budget," 12 April 2000. http://www.taxpayer.net/TCS/wastetbasket/transportation/2000-04-12bigdig.htm.
46. Cornelius Tacitus, *Annals of Imperial Rome—Book IV,* 109 A.D. Translated by Alfred John Church and William Jackson Brodribb. Classics.mit.edu/Tacitus/annals.11.iv.html.

47. Justin Sweet, *Legal Aspects of Architecture Engineering and the Construction Process,* third edition (St. Paul, Minnesota: West Publishing Co., 1985), 355.
48. Ibid.
49. Ibid., 105.
50. Robert F. Cushman and Christian G. Hedemann, *Architect and Engineer Liability: Claims Against Design Professionals: Second Edition, 1998 Cumulative Supplement* (New York: Aspen Publishers, Inc., 1998), 2.
51. This agreement clause is a slightly modified and shortened version of a contract clause that appears in an article written by Brian K. Stewart, Esq. and Christie Bodnar Swiss, Esq. entitled "How to Get a Design Professional to Work for Free!" Stewart and Swiss's clause is an example taken directly from an actual owner/architect agreement. In my clause I have substituted "design professional" for "architect." Stewart and Swiss's article is available online at http://www.aepronet.org/guest_essays.html.
52. Ibid.
53. Ibid.
54. The 9/11 World Trade Towers (WTTs) terrorist attack resulted in many more deaths than the Hyatt Regency disaster, but the WTTs' collapse was not caused by structural failure. It was caused by a chain reaction of events that eventually led to structural failure of the towers, hence the distinction.
55. DPIC Companies, Inc., *DPIC's Contract Guide: A Risk Management Handbook for Architectural, Engineering and Environmental Professionals* (Monterey, CA: DPIC Companies, Inc., 1999), III–37.
56. American Institute of Architects, *Standard Form of Agreement Between Owner and Architect with Standard Form of Architect's Services* (Washington, D.C.: American Institute of Architects, 1997), paragraph 2.6.2.1.
57. This agreement clause is a modified version of a contract clause that appears in Design Professionals Insurance Companies' (DPIC), *Contract Review and Revision 1: Contract Basics: An Education Course Based on DPIC's Contract Guide* (Monterey, CA: DPIC Companies, Inc., 2000), 29.
58. Ibid.
59. Ibid.
60. PSMJ Resources, Inc., Op Cit., 7.
61. Royal Barry Wills and Leon Keach, *The Business of Architecture* (New York: Reinhold Publishing Corporation, 1941), 174.
62. PSMJ Resources, Inc., Op Cit., 63.
63. Ibid., 61.
64. Ibid., 79.

Bibliography

The bibliography is an alphabetical listing, by author, of the various works consulted, quoted, or cited in this book. It includes books, articles, and Internet sources. The author does not claim that the bibliography is a comprehensive listing of publications on the subject of risk management nor is it necessarily a listing of all the works that influenced the text of this book.

American Institute of Architects. *The Architect's Handbook of Professional Practice.* Executive editor Joseph A. Demkin. 13th ed. New York: John Wiley & Sons, Inc., 2001.

American Institute of Architects. *The Business of Architecture: AIA Firm Survey.* Washington, D.C.: American Institute of Architects, 2006.

American Institute of Architects. *Standard Form of Agreement Between Owner and Architect with Standard Form of Architect's Services.* Washington, D.C.: American Institute of Architects, 1997.

Bernoulli, Jacob. *The Art of Conjecturing.* Originally published posthumously in 1713. Translation by Edith Dudley Sylla. *The Art of Conjecturing, Together with Letter to a Friend on Sets in Court Tennis.* Baltimore: The Johns Hopkins University Press, 2006.

Bernstein, Peter L. *Against the Gods: The Remarkable Story of Risk.* New York: John Wiley & Sons, Inc., 1998.

Bulfinch, Ellen Susan. *The Life and Letters of Charles Bulfinch, Architect with Other Family Papers.* Boston: Houghton, Mifflin and Company, 1896.

Cardano, Girolamo, translated by Jean Stoner. *The Book of My Life.* New York: New York Review of Books, 2002.

Cardano, Girolamo, translated by Sydney Henry Gould. *The Book on Games of Chance.* New York: Holt, Rinehart and Winston, 1961.

Coxe, Weld. *Marketing Architectural and Engineering Services.* New York: Van Nostrand Reinhold Company, 1971.

Cuff, Dana. *Architecture: The Story of Practice.* Cambridge, MA: The MIT Press, 1992.

Cushman, Robert F. and Hedemann, G. Christian. *Architect and Engineer Liability: Claims Against Design Professionals: Second Edition, 1998 Cumulative Supplement.* New York: Aspen Publishers, Inc., 1998.

DPIC Companies, Inc. *Contract Review and Revision 1: Contract Basics: An Education Course Based on DPIC's Contract Guide.* Monterey, CA: DPIC Companies, Inc., 2000.

DPIC Companies, Inc. *DPIC's Contract Guide: A Risk Management Handbook for Architectural, Engineering and Environmental Professionals.* Monterey, CA: DPIC Companies, Inc., 1999.

Fitch, James Marston. *American Building: The Historical Forces That Shaped It.* New York: Schocken Books, 1973. First published by Houghton Mifflin Company, 1947.

Getz, Lowell and Stasiowski, Frank. *Financial Management for the Design Professional: A Handbook for Architects, Engineers, and Interior Designers.* New York: Whitney Library of Design, 1984.

Hacking, Ian. *The Emergence of Probability: A Philosophical Study of Early Ideas about Probability Induction and Statistical Inference.* New York: Cambridge University Press, 1975.

Hacking, Ian. *The Taming of Chance.* New York: Cambridge University Press, 1990.

Harris, Joseph. *The Tallest Tower.* Bloomington, Indiana: Unlimited Publishing LLC, 2004.

Heller, Robert. *Making Decisions.* New York; DK Publishing, Inc., 1998.

Hitchcock, Henry-Russell. *The Architecture of H. H. Richardson and His Times.* Cambridge, MA: The M.I.T. Press, 1970. First published by the Museum of Modern Art, 1936.

Johnson, Paul-Alan. *The Theory of Architecture: Concepts, Themes and Practices.* New York: Van Nostrand Reinhold, 1994.

Kendrick, Tom. *Identifying and Managing Project Risk: Essential Tools for Failure-Proofing Your Project.* New York: AMACOM Books, 2003.

Kendrick, Tom. *The Project Management Tool Kit: 100 Tips and Techniques for Getting the Job Done Right.* New York: AMACOM Books, 2004.

LeGault, Michael R. *Think: Why Crucial Decisions Can't Be Made in the Blink of an Eye.* New York; Threshold Editions, 2006.

Levy, Matthys and Salvadori, Mario. *Why Buildings Fall Down: How Structures Fail.* New York: W. W. Norton & Company, 2002.

Meister, Maureen, editor. *H. H. Richardson: The Architect, His Peers, and Their Era.* Cambridge, MA: The MIT Press, 1999.

Morrison, Hugh. *Louis Sullivan: Prophet of Modern Architecture.* New York, W. W. Norton & Company, Inc. 1998.

Nelson, Charles. *Managing Quality in Architecture: A Handbook for Creators of the Built Environment.* Burlington, MA: Architectural Press, 2006.

Newton, Roger Hale. *Town & Davis Architects: Pioneers in American Revivalist Architecture 1812–1870 Including a Glimpse of Their Times and Their Contemporaries.* New York: Columbia University Press, 1942.

Pascal, Blaise, translated by A. J. Krailsheimer. *Pensées.* New York: Penguin Books, 1966.

Peirce Edition Project. *The Essential Peirce: Selected Philosophical Writings, Volume 2 1893–1913.* Bloomington and Indianapolis: Indiana University Press, 1998.

Pierson, William H., Jr. *American Buildings and Their Architects, Volume 1: The Colonial and Neoclassical Styles.* New York: Oxford University Press, 1970.

Pierson, William H., Jr. *American Buildings and Their Architects, Volume 2: Technology and the Picturesque, the Corporate and the Early Gothic Styles.* New York: Oxford University Press, 1978.

Piven, Peter and Perkins, Bradford. *Architect's Essential of Starting a Design Firm.* Hoboken, NJ: John Wiley & Sons, Inc., 2003.

PJMJ Resources, Inc. *A/E Financial Performance Survey,* 26th ed. 2006. Newton, MA: PSMJ Resources, Inc., 2006.

Project Management Institute. CD *A Guide to the Project Management Body of Knowledge,* third ed. (PMBOK). Newtown Square, PA: Project Management Institute, 2004.

Ramroth, William G., Jr. *Project Management for Design Professionals.* Chicago: Kaplan AEC Education, 2006.

Simmons, G. *Calculus Gems.* New York: McGraw Hill, 1992.

Stoppard, Tom. *Rosencrantz & Guildenstern Are Dead.* New York: Grove Weidenfeld, 1967.

Sweet, Justin. *Legal Aspects of Architecture Engineering and the Construction Process,* third ed. St. Paul, MN: West Publishing Co., 1985.

Van Rensselaer, Mariana Griswold. *Henry Hobson Richardson and His Works.* New York: Dover Publications, Inc., 1969. First published by Houghton, Mifflin and Company, 1888.

Wideman, R. Max, editor. *Project and Program Risk Management: A Guide to Managing Project Risks and Opportunities.* Newtown Square, PA: Project Management Institute, 1992.

Wills, Royal Barry and Keach, Leon. *The Business of Architecture.* New York: Reinhold Publishing Corporation, 1941.

Wintner, Steve L. and Tardif, Michael. *Financial Management for Design Professionals: The Path to Profitability.* Chicago: Kaplan AEC Education, 2007.

Index

A

Abduction, 29, 31
Accessibility requirements, 165
Accounting procedures, 78
Adam, Robert, *x*
Adler, Dankmar, 1, 4, 45
Administration support, 78
Admitted carriers, 181
Allen, Zachariah, 55, 56–58
American Arbitration Association (AAA), 202
American Beautiful Movement, 97
American Consulting Engineers Council (ACEC), 193
American Institute of Architects (AIA)
  dispute resolution, 202
  Document B141, 193, 221
  Document G813, 184–85
  ethics, 123
  founding of, 62–63
  standardized agreements, 193
American Society of Civil Engineers (ASCE), 193
Ames, Frederick Lothrop, 103
Ames II, Oliver, 103
Arbitration, 202
Architectural review boards, 165–67
Area summary calculation, 164–65
Aristotle, 7
Assumption, 42
Atwood, Charles, 96–97
Auditorium Building (Chicago), 1–2, 3–4

B

Beaux Arts Classicism, 44–45
Bell curve, 21
Benchmarks, 72–73
Bernoulli, Daniel, 24, 184
Bernoulli, Jacob, 17–20, 24
Bernoulli, Nicolaus, 17
Bernstein, Peter L., 15
Bid risk, 38
Bid runners, 149–50
Big Dig (Boston), 137, 151
Break-even multiplier, 237
Bronx–Whitestone Bridge (NYC), 127
Brunelleschi, Filippe, 131
Budget, 152–55
Building code, 160, 171–72
Building development review, 169
Bullfinch, Charles, *vii–xi, xiii–xiv,* 62, 203
Bullfinch, Ellen Susan, *ix*
Burnham, Daniel, 43–44, 96
Business development (BD)
  basics, 86–94
  chase anything that moves, 83–84
  checklist, 256–59
  clients, 89–94
  comprehensive risk management, 67
  firm organization, 86–87
  funnel analogy, 81–82
  long course, 100–102, 114
  plan, 94–100, 257
  project manager's role, 110–11
  projections, 97
  projects, 88–89
  requirements, 114–15
  risks, 37
  short course, 101, 102–4, 114
  silver bullet approach, 84
  silver spoon approach, 84–85
  strengths/weaknesses, 87–88
Business plan
  checklist, 256
  development, 69
  execution of, 72–73
  mission statement, 71–72, 73
  objectives, 72, 73
  organizational risk and, 66
  outline, 74
  purpose of, 70
  vision, 71, 73

C

Cambio, Arnolfo di, 131
Cardano, Girolamo, 6–9, 10
Career paths, 76
Certification clause, 196
Chargeability rate, 242–44
Chrysler Building, 160
City ordinance, 164
Claims-made basis policy, 183
Claims response plan
  comprehensive risk management, 63
  contractual risk, 187–91
  development, 70
  elements, 206
Client
  cost increases, 156–57
  existing, 106
  expectations/needs, 106–7
  financial stability of, 93
  payment history, 93–94
  pursuit of, 89–93
  risks, 38
  tracking, 99
Code analysis
  checklist, 264
  format, 162
  information, 163–65
  plan, 162
Codes, 160–69
Code summaries, 167–69
Columbian Exhibition (Chicago), 43–45, 96–97
Commodity-type client, 89–90
Communication skills, 123

Compensation demands, 190
Compensation procedures, 77
Competitively bid, 217–19
Comprehensive risk management
  checklist, 255
  plans, 65–70, 78
Connecticut State Capitol, 64
Construction budget
  checklist, 262
  designing to, 150–57
  owner's, 137
Construction management (CM), 219
Construction management at risk, 220
Construction phase risk, 39
Construction support services
  change orders, 225–26
  observation, 220–23
  procurement, 216–20
  project closeout, 228–30
  project delivery systems, 270
  request for pay, 233–34
  request for pay reviews, 226–28
  RFI responses, 223–25
  submittal review, 209–16, 270–72
Consultant
  checklist, 263–64
  purpose, 157–59
  risks, 38
  selection, 170–71
Contingencies, 135, 154, 155
Contingency planning, 42–43
Contractors, 149–50
Contractual risk, 38
Cost creep, 155
Cost plus fixed fee, 199
Cost plus fixed fee with guaranteed maximum, 199

D

Da Vinci, Leonardo, 126
Davis, Alexander Jackson, 61–62, 64–65, 96
De Moivre, Abraham, 20–23
Decision–making
  facts and, 24–25
  personal wealth and, 25
  potential rewards and, 14–16
  steps, 15
  utility and, 25
Deduction, 29, 31
Design budget, 157, 170, 263
Design/build
  guidelines, 234
  risk management, 230–31, 273
  support services, 220
Design phase completion, 248
Design team, 136–38
Developer, 93
Dewar, Donald, 129
Differential settlement, 2–4
Dispute resolution, 200–203, 206, 269
Documentation, 249

E

Egress requirements, 165
80/20 rule, 26–27, 53–54
Empire State Building, 160
Employee
  availability of, 107
  benefits package, 76–77
  career paths, 76
  compensation procedures, 77
  importance of, 72
  manual, 66, 69, 74–77
  professional development, 77
  recognition program, 76
  risks, 38
  turnover, 75–76
Employee relations and management plan. *See* Employee: manual
Employment policies, 75
Energy efficiency codes, 164
Engineers Joint Contract Document Committee (EJCDC), 193
Errors and omissions (E&O) insurance, 38, 177
Ethical behavior, 123
Excess and surplus (E&S) carriers, 181
Existing client, 106
Exiting requirements, 165
Expectations, 106–7, 133
Experienced client, 91
Explanation clause, 211, 212–13
External risk, 39

F

Factory Mutual, 58
Facts, 24–27, 39
Federal Customs House (NYC), 64
Federal Street Theater, *viii, ix*
Fermat, Pierre de, 11–12
Field inspectors, 222
Financial risk, 38, 39
Finch, James Marston, 79–80
Fire Marshal requirements, 164
First knowledge, 190–91
Fixed–price agreement, 197
Fluid mechanics, 24
FM Global, 58
Foundations, 2–4
Fractions, 5–9, 10–11, 30
Franklin Place (Boston), *ix*
Funnel analogy, 81, 82

G

Galileo, 10
Gardeners Cottage (North Easton), 104
Generalists, 121

Golden Gate Bridge, 127
Go/No Go process, 104–9, 257–58
Guarantee clause, 196

H

Hacking, Ian, 14–15
Harmonics, 127–28
Hazardous materials, 165
Height summary information, 165
Heller, Robert, 53–54
Histogram, 18, 19, 20–21
Hitchcock, Henry-Russell, 79, 80
Hollis Street Church (Boston), *viii*
Hunt, Richard Morris, 62–63
Huygens, Christiaan, 10
Hyatt Regency Hotel (Kansas City), 207–9

I

Incomplete drawings, 246–47
Indemnity clause, 196
Indiana State Capitol, 63
Induction, 29, 31
Injury, 190
Inspection, 221–22
Insurance company
- loss prevention, 55–58
- probability use by, 39

Insurance risk, 38

J

Johnson, Phillip, 85
Jouffret, Espirit Pascal, 21
Juran, Joseph, 26

L

Lambert, Phyllis, 85
Latrobe, Benjamin, *xiv*
Lawrie, Paul, 9–10
Leadership in Energy and Environmental Design (LEED), 165
Leaning Tower of Pisa, 2–3
Licensure, 138
Longwater Gate Lodge, 103
Loss prevention, 55–58
Louisiana State University, 28
Lump-sum agreement, 197

M

Manufacturers Mutual Fire Insurance Company, 57, 58
Marketing. *See also* Business development
- approach, 98–100
- costs, 108
- definition, 81–82
- fees, 108
- goals, 97–98
- plan, 67, 69
- procedures, 98–100

Massachusetts State House, *viii*
Matrix organization, 87
Mean, 21
Mediation, 202–3
Méré, Chevalier de, 10–11
Metropolis. Nicholas, 27
Miralles, Enric, 129
Mission statement, 71–72, 73, 94–95
Mitigation, 42
Modern architecture, 1, 79–80
Moisseiff, Leon, 127
Monte Carlo method, 27
Moral certainty, 19, 40
Morton Thiokol, 23

N

National Bureau of Standards, 208
National Society of Professional Engineers (NSPE), 123, 193
Needs, 106–7
Negotiated sole source, 216–17
Neuman, John von, 27
New client, 91
Newton, Roger Hale, 62
Newton, Sir Isaac, 20
New York Society of Architects, 62–63
Notice to proceed, 45
Novice client, 90–91
Nuisance claims, 191–92
Numerical probability, 48–51

O

Oakes Ames Memorial Hall, 103
Objectives, 72, 73, 128–29
Occurrence-basis policy, 182–83
Office procedures manual, 77–78
Office procedures plan, 66–67, 69
Office reporting and procedures plan. *See* Office procedures manual
Old Colony Railroad Station, 104
One standard deviation, 21–22
Organizational risk, 37, 66
Organizational structure, 95–97
Outcome predictibility, 20–23
Outliers, 22
Overall risk
- assessment matrix, 51
- determination of, 46–58
- factor, 47, 52–53
- management of, 47–48
- numerical probability, 48–51

P

Pareto, Vilfredo, 26
Paeto's principle, 26
Partnering, 200–201
Pascal, Blaise, 10, 11–12
Pascal's Wager, 12–13, 14–16
Pearson, Karl, 20
Peirce, Charles Sanders, 28–29
Permanent employer, 184
Personnel procedures, 78

Pierson, Jr., William H., *vii*, 103
Planning, 60
Planning ordinances, 160, 164
Portland Place, *x*
Practical certainty, 18–20
Pragmatic inference, 28–30, 31
Predictability range, 31
Prequalified contractor bids, 217
Prevention, 42
Pricing target multiplier, 237–38
Prior acts policy, 183
Private client, 92–93
Proactive risk management, 41–45, 59–60, 254–55
Probability
 calculation method, 7–9
 determination of, 35, 39–40
 expression of, 5–7, 30
 fractions and, 7–9, 10–11
Procurement services, 232
Product liability, 174–75, 265
Professional behavior, 123
Professional development, 77
Professional liability
 claims response plan, 187–91
 design professional, 175–76
 exposure management, 187
 nuisance claims, 191–92
 standard of care, 185–86, 205
Professional liability insurance
 claim response plan, 206
 claims, 267–68
 company agent, 179–80, 266
 deductibles, 178
 independent brokers, 179–80, 266
 legal defense, 178
 policies, 180–83, 205–6, 266–67
 purpose of, 177, 265
 risks, 38
Professional services contract
 agreement provisions, 193–95
 clauses, 203–5
 dispute resolution, 200–203
 exclusions, 199–200
 fee types, 196–99
 language, 195–96
 scope of services, 199–200
 standardized agreements, 192–93
Profitability rudders, 244–46
Project
 agreement clauses, 134
 budget, 136
 closeout, 228–30, 234
 comprehensive services, 134
 construction budget, 150–57
 contingencies, 135, 154, 155
 cost–control plan, 152–57
 cost creep, 155
 cost increases, 155–57
 delivery system, 216–20, 232, 270
 designer, 133
 events beyond your control, 129
 execution, 141–44
 failure, 125–31, 145
 financial performande, 143
 insurance, 178–79
 objectives, 128–29
 owner expectations, 133
 political sensitivity of, 133
 poor planning/execution, 131
 pre–sold, 105–6
 progress monitors, 155
 pursuit of, 88
 qualification, 107
 regulatory agency jurisdiction, 161
 reviews, 143–44
 risk, 38
 risk management, 133–34
 sale methods, 130
 schedule, 138
 scope creep, 154–55
 selection process, 107–8
 size, 88–89
 staff availability, 133
 tracking, 99, 104–5
 work plan, 139–41, 146
Project management
 checklist, 259–62
 manual, 66
 plan, 67, 69–70
 risks, 37
Project manager
 attitude, 121–22
 characteristics, 145–46
 communication, 124
 competency 120–21
 design team motivation, 124–25
 desire, 119
 ethical behavior, 123
 organizational skills, 119–20
 preventive risk management, 141–44
 project fee determination, 132–33
 risk management techniques, 132–41
 role of, 110–11
 selection, 132
 skills, 118–25, 259
Proposal, 258–59
Proposal request (PR), 225
Proposal writing, 111–13
PSMJ Resources, 238, 243
Public client, 92–93
Punch list, 230

Q

Qualitative analysis, 40
Qualitative probabilities, 59
Quality assurance plan, 67, 70
Quality control, review process flow diagram, 141
Quality risks, 38
Quantitative analysis, 39–40
Quantitative probabilities, 59

R

Rational conjecture, 17–18
Recovery planning, 41, 42, 43–44
Regulatory agency, 161
Repeat client, 91–92
Request for information (RFI)
  response, 223–25
  review procedures, 233
  risk management, 224–25, 272
Request for pay, 273
Request for quotation (RFQ)
  change orders, 225–26
  review procedures, 233
  risk management, 225–26, 272–73
Resources, 78
Retention, 227–28
Rewards
  decision–making and, 14–16
  risk proportionate to, 9–14
Richardson, Henry Hobson, 79–80, 96, 103–4
Risk
  acknowledgment of, 35, 36
  assessment matrix, 50, 51, 54–55
  assessment table, 49
  consequences, 35
  definition of, *xi*
  determination of, 35, 36–39
  impact, 40–41, 51–53
  proposal writing and, 111–13
  and reward, 9–14, 30
Risk management
  definition of, *xii*
  fractions, 5–7
  Pascal's Wager, 12
  pragmatic inference and, 28–30
  principles, 251–52
  procedures/solutions, 35
  process, 2, 4–5, 253–55
  solutions/procedures, 45–46
  steps, 35–46, 58–59
  techniques, 132–41
  tools, 5–9, 67–68
Risk manager, 125
Risk–source categories, 35, 37–39, 59, 253–54
Rohe, Mies van der, 85
Root, John, 96
Royal Crescent (England), *ix*
Rule of thumb
  design phase completion, 248
  document, 249
  don't panic, 241–42
  incomplete drawings, 246–47
  profitability rudders, 244–46
  scope creep management, 239–41
  score keeping, 246
  60 percent rule, 242–44
  territory, 236–37
  3.0 multiplier, 237–38

S

Sales, 81, 82
Samplings, 17–20, 20–25
San Francisco–Oakland Bay Bridge, 127
Santa Maria Dome (Italy), 131
Scope creep, 154–55, 239–41
Scope of work, 135
Score keeping, 246
Scottish Parliament Building
  Blast Wall, 128
  costs, 90, 129, 151
Seagram Building (NYC), 85
Selection panel, 108
Seller–and–doer model, 95–96
Seller/doer model, 95–96
Shop drawing log, 211
Shop drawing stamp, 211
60 percent rule, 242–44
Slow burning construction, 57–58
Special governmental agencies/districts, 164
Standard deviation, 21–22
Standard Form of Agreement Between Owner and Architect, 193, 202
Standard Form of Agreement Between Owner and Engineer, 193
Standard Form of Agreement for the Temporary Placement of Employees, 184–85
Standard Form of Architect's Services: Design and Contract Administration, 221
Standard of care, 185–86, 205
Strengths/weaknesses, 87–88
Studios, 87
Submittal conformation procedures, 211
Submittal coordinator, 210–11
Submittal rejections, 214–15
Submittal review
  checklist, 270–72
  guidelines, 213–14
  procedures, 232

process, 212, 213
support services, 209–16
transmittal letter, 211
Sullivan, Louis, 1, 44, 45, 79
Suspension bridge, 127
Sweet, Justin, 176
Sydney Opera House, 90, 151

T

Tacitus, Cornelius, 160
Tacoma Narrows Bridge (Washington), 126, 127
Tailor's Rule of Thumb, 235
Technical performance, 37
Technological risk, 38
Teller, Edward, 27
Temporary employer, 184
Territory, 236–37
Textile mills, 56–57
3.0 multiplier, 237–38
Three standard deviation, 22
Time and materials agreement, 130, 197–98
Time and materials with maximum-not-to-exceed, 198–99
Toutine Crescent (Boston), *ix*
Town, Ithiel, 61–62, 64–65, 96
Traditional architecture, 1, 80
Transference, 42
Transportation Building, 44, 45
Trinity Church (Boston), 102, 103
Two standard deviation, 22

U

Ulam, Stanislaw, 27–28
U.S. Capitol Building, *xiii, xiv*
Utility, 24–27, 30
Utilization rate, 242–44

V

Value-added client, 89–90
Values, 72
Van de Velde, Jean, 9–10
Vision, 71, 73
Vitruvius, 2

W

Warranty clause, 196
What if modeling, 27
Woods, Tiger, 9, 10
Work breakdown structures (WBS), 139
Work plan, 139–41, 146
Work product standards, 38
Wrongdoing, 190